Chemistry Research and Applications

nova
science publishers
New York

Chemistry Research and Applications

Cyanide: Occurrence, Applications and Toxicity
Bill M. Torres (Editor)
2022. ISBN: 978-1-68507-619-1 (Softcover)
2022. ISBN: 978-1-68507-670-2 (eBook)

A Closer Look at Carvacrol
Zak A. Cunningham (Editor)
2022. ISBN: 978-1-68507-627-6 (Softcover)
2022. ISBN: 978-1-68507-634-4 (eBook)

Fundamentals of Photocatalysis
Orva Auger (Editor)
2021. ISBN: 978-1-68507-374-9 (Softcover)
2021. ISBN: 978-1-68507-417-3 (eBook)

Polypropylene: Advances in Research and Applications
Théodore Marleau (Editor)
2021. ISBN: 978-1-68507-378-7 (Hardcover)
2021. ISBN: 978-1-68507-401-2 (eBook)

Boron: Advances in Research and Applications
Lynn Mcconnell (Editor)
2021. ISBN: 978-1-68507-231-5 (Softcover)
2021. ISBN: 978-1-68507-259-9978 (eBook)

Applications of Layered Double Hydroxides
Rajib Lochan Goswamee, PhD (Editor)
Pinky Saikia, PhD (Editor)
2021. ISBN: 978-1-68507-355-8 (Hardcover)
2021. ISBN: 978-1-68507-381-7 (eBook)

More information about this series can be found at
https://novapublishers.com/product-category/series/chemistry-research-and-applications/

Bill M. Torres
Editor

Cyanide

Occurrence, Applications and Toxicity

nova science publishers
New York

Library of Congress Cataloging-in-Publication Data

ISBN: 978-1-68507-619-1

Published by Nova Science Publishers, Inc. † New York

Contents

Preface

While cyanide finds application in myriad industries, its presence in the environment as a pollutant warrants special attention, as exposure to cyanide is harmful or even lethal to humans and animals. This volume includes five chapters that examine cyanide's industrial applications and methods for reducing its environmental impact. Chapter one introduces electrochemical sensors based on silver and silver sulfide nanoparticles supported on multi-walled carbon nanotubes and hierarchical porous carbon for the detection of low concentrations of free cyanide in water. Chapter two reviews theoretically predicted cyanide compounds having strong bonding units which can be synthesized under roughly ambient conditions. Chapter three proposes a parsimonious global cyanide cycle to minimize its presence in the environment. Chapter four describes commercially available methods for cyanide removal including reverse osmosis, chemical oxidation and biological treatment, focusing on Moving Bed Biofilm Reactor technology. Lastly, chapter five examines the efficacy of Fusarium oxysporum for microbial degradation of cyanide to treat wastewater by measuring the heat capacity of its lyophilized biomass using a modulated differential scanning calorimeter.

Chapter 1 - The environmental pollution derived from the use of cyanide in mineral extraction has triggered numerous social conflicts in developing countries. Particularly, the free cyanide (CN-) found in mining effluents requires regular monitoring given its high toxicity. However, the expensive instrumental techniques for its analysis do not provide results in a short time. For this reason, having an electrochemical sensor capable of analyzing CN- would significantly contribute to its regulation, ultimately generating a better population-mining relationship. In this chapter, electrochemical sensors based on silver (Ag-NP) and silver sulfide nanoparticles (Ag2S-NP), supported on multi-walled carbon nanotubes and hierarchical porous carbon, are introduced. The Ag-NP and Ag2S-NP offer good selectivity for free cyanide detection, and the carbonaceous materials (CMs) employed as support stabilize the nanoparticles, improving their selectivity and sensitivity. A layer-by-layer self-assembly method was used to obtain the Ag-NP/CM, while an in-situ co-precipitation method was invoked to obtain the AgS2-NP/CM. The electrochemical sensors exhibit detection limit values ranging from 1.75 μg L-1 to 9.22 μg L-1 for samples from mining effluents, i.e., within the international standards of 200 μg L-1. Selectivity coefficient values of about 1 with respect to interfering ions such as: chlorides, sulfides, and carbonates, demonstrate the specific response of these materials toward CN-. In summary, this chapter presents several physicochemical characteristics showing the synergy between carbon materials with Ag-NP and Ag2S-NP for the detection of low concentration of CN- in water. Finally, given to the good characteristics that these electrochemical sensors exhibit, they emerge as a good alternative for monitoring cyanide, thus helping to solve social problems.

Chapter 2 - Compounds containing cyanide (–CN) and isocyanide (–NC) functional groups are of utmost importance in chemistry owing to their prominent astronomical applications including their extensive involvement in star-formation, cold clouds formation, and in circumstellar shells. More than 200 molecules having –CN functional group are detected in interstellar medium and around 40 of them are synthesized in the laboratory as well. A number of alkali, alkaline earth, and transition metal containing cyanides and isocyanides such as NaCN, KCN, MgCN, MgNC, CaNC, AlNC, SiCN, SiNC, FeCN, FeNC and a list of other interstellar species including more complex molecules like isopropylcyanide are detected astronomically. In this mini-review, the authors briefly epitomize their works to date on theoretically predicted cyanide compounds having strong bonding units which can be synthesized under roughly ambient conditions. The authors' studied complexes include MNgCN/MNgNC, LMCN/LMNC, and NCNgNSi where Ng = Xe, Rn; M = Cu, Ag, Au; and L = C2H2, C2H4, CO, N2, NH3, H2O, H2S, and 1,3-dimethylimidazole (DMI). They have thermochemical stability concerning most of the dissociation paths, and their kinetic stability with respect to those channels which are thermodynamically spontaneous indicate their possible experimental realization. In combination with the structure and stability, the authors' quantum chemical study also discusses the bonding senario in these compounds with the help of standard theoretical techniques like natural bond orbital, energy decomposition and electron density analyses. The authors also carefully examine and discuss possible ways of trapping the transient metal isocyanide isomer by introducing suitable ligands in an attempt to increase their kinetic stability. This report on cyanide and isocyanide compounds from the perspective of density functional theory along with high-level computations using coupled-cluster theory will guide experimentalists with information on structures, energetics, stability and, nature of chemical bonding which are essentially unknown.

Chapter 3 - Cyanide (CN-), a highly toxic anion, has drawn special interest due to its affection for many biological functions and wide applications in numerous chemical processes. To regulate cyanide in drinking water, various countries have set regulations on the maximum contaminant level for cyanide in water resources. Cyanide is one of the toxic and environmental inorganic pollutants. Due to the extreme toxicity of cyanide ions, the world health organization (WHO) recommended a tolerable limit of cyanide ion at 1.9 μM, while US environmental protection agency recommends the cyanide ion limit at 7.8 μM. Cyanide toxicity occurs because this compound strongly binds to metals, inactivating metalloenzymes such as cytochrome c-oxidase. Cyanide blocks the oxidative respiration pathway, impeding oxygen usage within tissues; the major metabolic pathway results in the formation of less toxic thiocyanate. Cyanogen chloride may be formed when cyanide polluted water is treated by chlorination. A parsimonious global cyanide cycle is proposed in this Chapter.

Chapter 4 - Cyanide leaching is the most commonly used method for gold extraction. Although cyanide recovery and destruction is undertaken, excess water still contains cyanide residual and cyanide compounds such as cyanate and thiocyanate. These compounds are toxic to aquatic life and need to be removed from water before discharge to the environment to comply with local regulations. Commercially available methods for cyanide removal will be described including reverse osmosis, chemical oxidation and biological treatment. A focus on biological treatment using Moving Bed Biofilm Reactor (MBBR) technology will be made. Extensive testing at a mine site in Ghana has demonstrated that biological treatment is a cost effective and sustainable solution for cyanide removal from mine effluents.

Chapter 5 - Cyanide is a well-known constituent of mine wastewater that can be degraded by various processes. However, due to the cost and environmental challenges, microbial degradation seems to be the most effective process. When wastewater is treated with microorganisms, process performance should not only be based on toxicant degradation but also the impact of the toxicant on the physical properties of the microorganisms. The heat capacity of lyophilised biomass of Fusarium oxysporum was measured using modulated differential scanning calorimeter. The heat capacities for F. oxysporum grown in cyanidation wastewater were 1.1982, 1.077 and 1.143 J K-1 g-1 on glucose (GA), Beta vulgaris (BA) and cyanide supplemented with Beta vulgaris (BCN), respectively at 298.15 K and 1 atm. The enthalpies of formation of dry biomass are -297.58, -233.07 and -278.60 kJ/C-mol for BA, BCN and GA, respectively. These values were found to be within the range of some biological molecules. The presence of cyanide in the wastewater minimally affected the thermodynamic property of the dried biomass of F. oxysporum.

Chapter 1

Electrochemical Sensors Based on Carbonaceous Materials Modified with Silver Nanoparticles and Silver Sulfide Nanoparticles for the Detection of Low Concentrations of Free Cyanide

Andy A. Cardenas-Riojas[1], Ademar Wong[2,3], Golfer Muedas-Taipe[1], Adolfo La Rosa-Toro[1], Maria D.P.T. Sotomayor[2,3], Miguel Ponce-Vargas[4] and Angélica M. Baena-Moncada[1,*]

[1]Laboratorio de Investigación de Electroquímica Aplicada,
Facultad de Ciencias de la Universidad Nacional de Ingeniería, Av. Túpac Amaru 210,
Rímac, Lima, Perú
[2]Department of Analytical Chemistry, Institute of Chemistry,
State University of São Paulo (UNESP), Araraquara, SP, Brazil.
[3]National Institute for Alternative Technologies of Detection,
Toxicological Evaluation and Removal of Micropollutants
and Radioactives (INCT-DATREM),
Araraquara, SP, Brazil
[4]Institut de Chimie Moléculaire de Reims, UMR CNRS 7312,
University of Reims Champagne-Ardenne, Moulin de la Housse,
Reims, France

---

[*] Corresponding Author's E-mail: abaenam@uni.edu.pe.

In: Cyanide: Occurrence, Applications and Toxicity
Editor: Bill M. Torres
ISBN: 978-1-68507-619-1

Abstract

The environmental pollution derived from the use of cyanide in mineral extraction has triggered numerous social conflicts in developing countries. Particularly, the free cyanide (CN^-) found in mining effluents requires regular monitoring given its high toxicity. However, the expensive instrumental techniques for its analysis do not provide results in a short time. For this reason, having an electrochemical sensor capable of analyzing CN^- would significantly contribute to its regulation, ultimately generating a better population-mining relationship. In this chapter, electrochemical sensors based on silver (Ag-NP) and silver sulfide nanoparticles (Ag_2S-NP), supported on multi-walled carbon nanotubes and hierarchical porous carbon, are introduced. The Ag-NP and Ag_2S-NP offer good selectivity for free cyanide detection, and the carbonaceous materials (CMs) employed as support stabilize the nanoparticles, improving their selectivity and sensitivity. A layer-by-layer self-assembly method was used to obtain the Ag-NP/CM, while an *in-situ* co-precipitation method was invoked to obtain the AgS_2-NP/CM. The electrochemical sensors exhibit detection limit values ranging from 1.75 μg L^{-1} to 9.22 μg L^{-1} for samples from mining effluents, i.e., within the international standards of 200 μg L^{-1}. Selectivity coefficient values of about 1 with respect to interfering ions such as: chlorides, sulfides, and carbonates, demonstrate the specific response of these materials toward CN^-. In summary, this chapter presents several physicochemical characteristics showing the synergy between carbon materials with Ag-NP and Ag_2S-NP for the detection of low concentration of CN^- in water. Finally, given to the good characteristics that these electrochemical sensors exhibit, they emerge as a good alternative for monitoring cyanide, thus helping to solve social problems.

Keywords: free cyanide, electrochemical sensors, carbonaceous materials (CMs), layer-by-layer self-assembly, silver nanoparticles (Ag-NP) and silver sulfide nanoparticles (Ag_2S-NP)

1. Introduction

Cyanides are chemical species containing a carbon atom triple-bonded to a nitrogen atom (cyano group), being highly reactive and lethal to humans. For centuries cyanide was used as poison, and it was only isolated in 1782 by the Swedish chemist Scheel (Bläsing et al., 2020). Cyanide is most commonly found in the environment in its free form (CN^-), as hydrocyanic acid (HCN), or as water-soluble alkaline salts, such as sodium cyanide (NaCN) and potassium cyanide (KCN) (Lovasoa et al., 2017; Luque-Almagro et al., 2011). It can be produced by microorganisms such as algae, bacteria and fungi and can be found in plants, foods and -in low concentrations- in soil and water (Jaszczak, Polkowska, et al., 2017; Logue et al., 2010). Plants such as almonds, cassava, flax seeds and bamboo shoots are considered toxic if contain cyanide at a concentration of 200 ppm in the form of cyanogenic glycosides (cyanide releasing plants) (Bolarinwa et al., 2016; Figueira et al., 2016; Jaszczak, Polkowska, et al., 2017). The cyanide ion is able to bind strongly to ferric ions in the mitochondrial respiratory chain cytochrome oxidase, reducing enzyme activity by blocking oxygen transport to cells. This results in the interruption of energy production by organism (Leavesley et al., 2008; Manoj et al., 2020). In general, cyanide is toxic in its various forms to terrestrial, aquatic and aerial organisms and acts as an inhibitor of metalloenzymes. The generation of effluents by industries is the main source of release of complex cyanides into the environment. The applicability of cyanides is made in large scale for a wide variety of industries such as in the metallurgy, electroplating, mining, polymer synthesis, agrochemicals, pharmaceutical compounds and dyes industries (Alvillo-Rivera et al., 2021). Effluents from mining and agribusiness are considered a major source of cyanide contamination in the environment. One form of cyanide poisoning occurs mainly through ingestion of food which is in turn absorbed from the gastrointestinal tract and distributed throughout the body (Dash et al., 2009; Egekeze & Oehme, 1980). The most serious consequence of cyanide ingestion is the blockage of the respiratory chain and the inhibition of oxygen metabolism, while the effects of acute exposure mainly occur in the central nervous and cardiovascular systems (Hamel, 2011). Initial symptoms include headache, dizziness, shortness of breath and vomiting. It can be followed by seizures, low blood pressure, loss of consciousness and cardiac arrest (Hamel, 2011; Parker-Cote et al., 2018). A concentration between 0.5 and 3.5 mg per kg is generally lethal to humans (Singh et al., 2006). The exposed person, if survives, will probably have neurological problems. It should be remembered that the toxicodynamic effects of cyanide may vary with dose, rate of administration, chemical form of the cyanide, and other factors, including gender, age, weight, and physical condition. Rat studies showed that the cyanide can kill in seconds to a few hours depending on cyanide exposure levels (Lawson-Smith et al., 2011; Rice et al., 2018).

Cyanide ion is the most toxic form of this compound due to its powerful tissue action, rapid absorption and circulation in the blood. For this reason the

World Health Organization (WHO) has established a maximum acceptable level of cyanide in drinking water of 1.9 μmol L^{-1} (Hernández et al., 2017). In this regard, the spill of millions of liters of cyanide waste in Romania in 2000 (Zelder & Männel-Croisé, 2009) represents one of the worst cases of water contamination in Europe.

An efficient monitoring of the residual cyanide ion concentration is crucial due to its extreme toxicity. The choice of method for cyanide detection will depend on the type of matrix and the objectives of the experiment. Simpler experimental methods like turbidity due to the formation of a precipitate by an argentometric titration, and more sophisticated instrumental methods such as chromatography have been used for determination of cyanide ion in different matrices. Other analytical methods like colorimetry (Bhaskar & Sarveswari, 2019; Erdemir & Malkondu, 2020; Ou et al., 2015; Sasikumar & Ilanchelian, 2020), fluorometry (Agarwalla et al., 2015; Long et al., 2019; Manickam & Iyer, 2020; Y. Yu et al., 2018), optical (Afkhami & Sarlak, 2007; Nandi et al., 2017), capillary electrophoresis (Chinaka et al., 2001; Q. Zhang et al., 2015), electrochemical (Cárdenas Riojas et al., 2019; Carter & Rimmer, 2004; Ghanavati et al., 2014; Lindsay & O'Hare, 2006a), spectrophotometric (Alizadeh et al., 2016; Gopal Reddy et al., 2010; Surleva et al., 2013), titration(Breuer et al., 2011; Suzuki et al., 2003) and chemiluminescence (Lu et al., 1995; Lv et al., 2005) can also be mentioned (Table 1).

In the literature is possible to find different reagents that were used to determination of cyanide. Among them, we have the methyl violet reagent that reacts with cyanide ion decreasing the absorbance (Afkhami & Sarlak, 2007) and reagents that can change its color like ninhydrin (Surleva et al., 2013). Both systems can be used to detect cyanide. The use of sensors is suitable by using specific reagents that are generally based on irreversible reactions. In this case, the presence of cyanide ion is signaled through a change in color or fluorescence (Z. Xu et al., 2010). The use of fluorometric assays based on optical probes has some advantages such as high selectivity, simplicity of implementation, fast response times, high sensitivity, enabling the use of this technique in medical centers.

Electrochemical methods have also been highlighted, especially with electrochemical sensors due to their numerous advantages such as high sensitivity and selectivity using simple and low-cost electrochemical platforms (Dhahi et al., 2010; Stradiotto et al., 2003). The possibility of direct analysis without a pre-treatment step and the numerous analyzes that can be performed, as well as the wide applicability in different samples make the electrochemical methods highly promising.

Table 1. Analytical parameters obtained from different methodologies for cyanide determination in different matrices

Methodology	Solvent/solution	pH	Linear range (µmol L^{-1})	LOD (µmol L^{-1}) or signal	Reference
Colorimetric and fluorogenic	CH_3CN/H_2O (9/1)	7.0	-	0.45	(Erdemir & Malkondu, 2020)
Colorimetric	H_2O	<8	-	0.2	(Bhaskar & Sarveswari, 2019)
Colorimetric	Au NBPs	5.0	1 - 15	1.6×10^{-3}	(Sasikumar & Ilanchelian, 2020)
Colorimetric	Acetonitrile–water (95:5, v/v)	-	-	8.0	(Ou et al., 2015)
Chromatography with electrochemical detection	NaOH	12	0.01 - 5.8	3.8×10^{-3}	(Jaszczak, Narkowicz, et al., 2017)
Chromatography with electrochemical detection	$Na_2B_4O_7 \cdot 10H_2O$	-	0.038 – 38.5	0.038	(Koch, 1983)
Fluorescent	CH_3CN and $C_{16}H_{36}F_6NP$	7.0	-	39×10^{-3}	(Manickam & Iyer, 2020)
Fluorescent	Potassium phosphate buffer/ dimethylformamide (4:6, v/v)	7.4	0 - 60	0.23	(Long et al., 2019)
Fluorescent	$EtOH/H_2O$ (v/v = 1/1)	-	-	0.059	(Y. Yu et al., 2018)
Fluorescent	HEPES/CTAB	7.2	-	0.28	(Agarwalla et al., 2015)
Optical sensor	Phosphate buffer	7.0	3.8×10^{3} - 95×10^{3}	2.4×10^{3}	(Afkhami & Sarlak, 2007)
Optical sensor	H_2O	8.0	-	9.4	(Nandi et al., 2017)
Capillary electrophoresis with fluorescence detection	Borate buffer and KCN	9.2	-	4.0×10^{-3}	(Q. Zhang et al., 2015)
Capillary electrophoresis with fluorescence detection	Borax buffer and methanol (8:2, v/v)	6.0	0.0038 – 7.7	3.8×10^{-3}	(Chinaka et al., 2001)
Electrochemical sensor	Phosphate buffer	11.4	-	4.0	(Lindsay & O'Hare, 2006b)
Electrochemical sensor	NaOH	12	0.59 - 1100	0.07	(Cárdenas Riojas et al., 2019)

Table 1. (Continued)

Methodology	**Solvent/solution**	**pH**	**Linear range (μmol L^{-1})**	**LOD (μmol L^{-1}) or signal**	**Reference**
Potenciometric sensor	NaOH	12	-	0.12	(A. A. Cárdenas-Riojas et al., 2019)
Electrochemical biosensor	Phosphate buffer	7.5	1.6 - 13	0.43	(Ghanavati et al., 2014)
Spectroscopic with chromofluorogenic detection	THF/H_2O solution (1:1 v/v, pH = 7.15)	7.15	-	0.81	(Alizadeh et al., 2016)
Spectrophoto-metric	Na_2CO_3 + NaOH	10.8	-	0.38	(Surleva et al., 2013)
Raman spectroscopy	H_2O	-	-	3.8	(Gopal Reddy et al., 2010)
Titration	H_2O	12.6	-	color	(Suzuki et al., 2003)
Titration	H_2O	12	-	precipitation	(Breuer et al., 2011)
Chemi-luminescence	NaOH	-	0.5 - 50	0.23	(Lv et al., 2005)
Chemi-luminescence	NaOH	-	0.19 – 0.76	0.19	(Lu et al., 1995)

To determine cyanide ion in aqueous media by electrochemical method, voltammetric techniques as cyclic voltammetry and pulse voltammetry (differential pulse voltammetry and square wave voltammetry) had been widely used in the development of electrochemical sensors (Cárdenas Riojas et al., 2019; Chinaka et al., 2001; Lindsay & O'Hare, 2006b). Potentiometric method has also been used successfully by ion selective electrodes (ISEs) (A. A. Cárdenas-Riojas et al., 2019).

It should be noted that the methodology to be used may vary according to the nature of the sample, the legislation of each country, the instrumentation available, specialized technical, the concentration that is intended to be determined and the type of sample to be analyzed.

As far as we know, cyanide-based sensors have not been completely overhauled in recent years. This chapter discusses the utilization, contamination, and toxicity of cyanide. At first, a general approach to the most used methodologies for detection of cyanide ion was presented. Then, a specific approach was carried out using carbonaceous materials modified with silver nanoparticles and silver sulfide nanoparticles.

2. Sensor Based on Silver Nanoparticles Supported on Carbon Materials for the Potentiometric Detection of Free Cyanide

There are various methods to support metallic nanoparticles on carbon materials. One method is the direct chemical reduction of the precursor onto the substrate. However, this synthesis method does not generate monodisperse nanoparticles, which are highly dependent on the diffusion of the precursor and the reducing agent into the carbon structure (Wu et al., 2011). Another method consists of synthesizing the nanoparticles by some "bottom up" method with their subsequent anchoring in the carbons structure using the self-assembled layer-by-layer technique. The synthesis of the nanoparticles by the bottom-up method allows to obtain monodisperse nanoparticles that are physically the same (crystalline structure, morphology, size, etc.) (Podsiadlo, 2008). The layer-by-layer self-assembly method is widely used for the construction of nanostructured materials, such as thin films on a nanometric scale, allowing the incorporation of high amounts of nanoparticles (Decher & Hong, 1991). The formation of electrostatically self-assembled multilayers was proposed by Decher in 1991. The layer-by-layer self-assembly technique is based on the sequential adsorption of layers of opposite charge on a surface by electrostatic interaction (Decher & Hong, 1991). Based on this, 3 different types of systems can be assembled:

(i) Polymer-Polymer
(ii) Polymer-Nanoparticle
(iii) Nanoparticle-Nanoparticle.

A cycle of the deposition process in the layer-by-layer self-assembly is shown in Figure 1. In the first step, the substrate (in this case glass) is immersed in a solution containing the component of opposite charge, generally charged polyelectrolytes are used to allow the coverage of a high surface area on the substrate. In step two the substrate is immersed in the solvent to remove polyelectrolyte excess and what is weakly bound to the substrate. Subsequently, the substrate is immersed in a component that has the opposite charge to the polyelectrolyte used in point 1 (charged nanoparticles or polyelectrolyte), and then immersed again in the solvent. Once this cycle is finished, a bilayer is obtained. This process can be repeated several times until the desired number of layers is obtained (Decher & Hong, 1991).

Polyelectrolytes (polycations and polyanions) are polymers that consist of ionizable groups attached to charged polymer chains. They show a strong tendency to adsorb. A widely used polyelectrolyte is poly (diallyldimethylammonium) chloride (PDADMAC), the structure of the polyelectrolyte can be seen in Figure 1b. The layer-by-layer self-assembly technique allows to support silver nanoparticles (Ag-NP) in different types of carbons: multi-walled carbon nanotubes (MWCNT), and Hierarchical Porous Carbon (HPC).

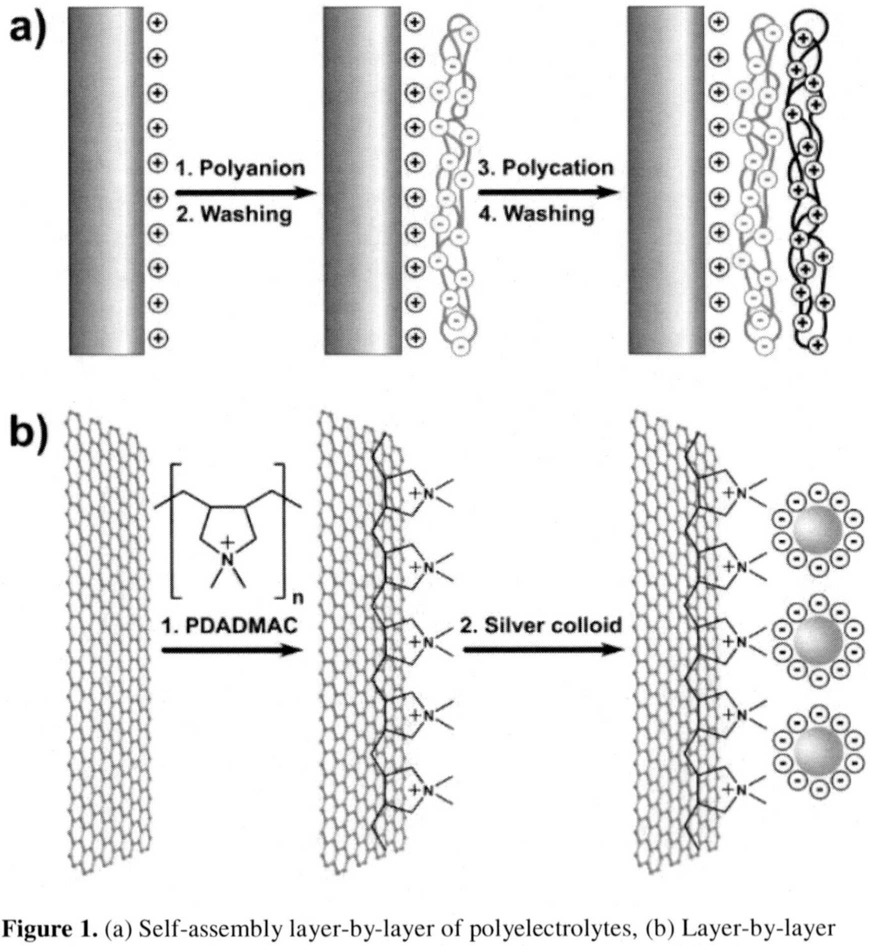

Figure 1. (a) Self-assembly layer-by-layer of polyelectrolytes, (b) Layer-by-layer self-assembly technique employed to support silver nanoparticles (Ag-NP) with Poly (diallyldimethylammonium chloride), PDADMAC in CMs. (Alexeyeva & Tammeveski, 2008).

In the last decades, Ag-NP have been used as electrochemical sensors (Hongyuan Zhang et al., 2020a). In this particular case, Ag-NP are used for the electrochemical detection of free cyanide, because the Ag-NP forms a complex with cyanide at basic pH (HISKEY & ATLURI, 1988). In aqueous solution, Ag (I) forms a series of cyanide ions complexes: $[Ag(CN)_2]^-$, $[Ag(CN)_3]^{2-}$ and $[Ag(CN)_4]^{3-}$ (Dehnicke, 1976). The $[Ag(CN)_2]^-$ is the predominant species, in addition to that, dilute cyanide solutions are normally used during cyanidation (reactions 1 and 2). (Dehnicke, 1976)

$$Ag^+ + 2CN^- \leftrightharpoons [Ag(CN)_2]^-, \beta = 2.8x10^{20} \qquad \text{reaction 1}$$

$$[Ag(CN)_2]^- + CN^- \leftrightharpoons [Ag(CN)_3]^{2-}, K = 35.5 \qquad \text{reaction 2}$$

Moreover, in alkaline cyanide solutions at pH values above 10, silver oxidizes to $[Ag(CN)_2]^-$ at approximately - 0.19 V in CN^- $1x10^{-4}$ mol L^{-1}. However, below pH 4 $[Ag(CN)_2]^-$ reactions with hydrogen ions to form a AgCN solid phase. Under acidic conditions (pH < 2.6), the $AgCN_{(s)}$ phase dissolves to produce Ag^+ and HCN (Dehnicke, 1976).

In recent times, many methods have been designed to synthesize nanoparticles. The assembly of nanoparticles would benefit from the development of clean, non-toxic and environmentally acceptable chemical methods (Han Zhang et al., 2016). Involving the reduction of a silver salt such as silver nitrate by adding a reducing agent (sodium borohydride, ascorbic acid, hydrazine, citric acid, amino acids and carbohydrates, etc.) (Zaheer & Rafiuddin, 2012), in the presence of a suitable stabilizer (surfactants, polymers, triblock polymers, gelatin, phospholipids, dendrimers and cellulose)(Zaheer & Rafiuddin, 2012). In this work, Ag-NP were synthetized using tyrosine as reducing agent. The procedure used was described by (Selvakannan et al., 2004). For this, 10.0 mL of $AgNO_3$ 1.0 mmol L^{-1} was mixed with 10.0 mL of tyrosine1.0 mmol L^{-1}, in a 250 mL Erlenmeyer flask, and stirred. 80 mL of ultrapure water was added to the mixture and 1 mL of KOH 0.1 mol L^{-1} was added to achieve a pH = 10. Then, the mixture was heated until the formation of a yellow colloidal solution, indicating the formation of Ag-NP. The Ag-NP were centrifuged at 12000 rpm for 5 min, the supernatant solution was removed, and ultra-pure water was added at pH 10, the process was repeated 3 times.

The amino acid tyrosine is an excellent reducing agent under alkaline conditions and can be used to reduce Ag^+ ions to obtain stable Ag-NP in water. Tyrosine-Ag-NP can be separated into a precipitate that is easily redispersed in water (Zaheer & Rafiuddin, 2012). The reduction of silver ions at high pH occurs due to the ionization of tyrosine phenolic group which is then capable of reducing Ag^+ ions and converted to a semi-quinone structure (Zaheer & Rafiuddin, 2012). Reaction 3 shows the Ag-NP formation mechanism using tyrosine as a reducing agent: in the reaction (pH = 7.9), tyrosine forms a complex with Ag^+ ions. The following reaction shows the electrons transfer from the tyrosine phenolic group to Ag^+ leading to the formation of Ag^0 and tyrosyl radical (Zaheer & Rafiuddin, 2012).

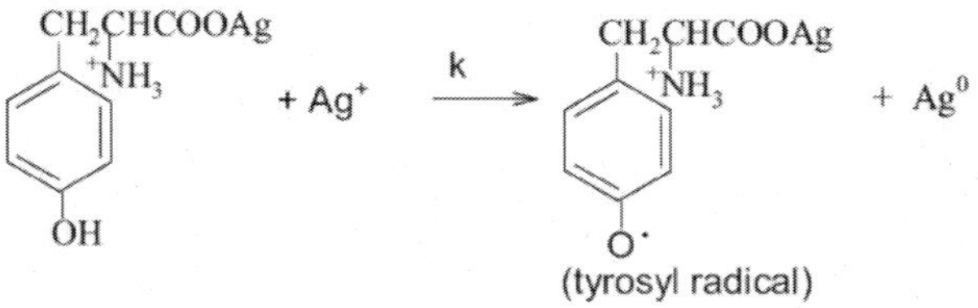

reaction 3

The average sizes of the Ag-NP were obtained by Dynamic Light Scattering (DLS) analysis, an effective diameter of 57 nm was obtained and a polydispersity of 0.300. In Figure 2a the UV-vis analysis is shown. Ag-NP presents an absorption band at 413 nm; which indicates the presence of Ag-NP at nanometric scale near to the theoretical value (400 nm), in addition to the characteristic yellow color of the colloid (Cruz et al., n.d.). The electrochemical behavior of the synthesized Ag-NP was studied by cyclic voltammetry (CV). To perform the experiment, 20 μL of Ag-NP colloid were deposited in a glassy carbon electrode (working electrode), using Ag/AgCl 3.0 mol L^{-1} as reference electrode and graphite bar as auxiliary electrode. Figure 2b presents the CV of Ag-NP in KOH 0.1 mol L^{-1} as supporting electrolyte at 0.05 V s^{-1}, obtaining the oxidation and reduction potentials characteristic of silver; where the oxidation potential is 0.36 V and the reduction potential is 0.05 V (Cloake et al., 2015; Jones et al., 2008). The synthesized Ag-NP were supported on HPC and MWCNT by layer-by-layer self-assembly method.

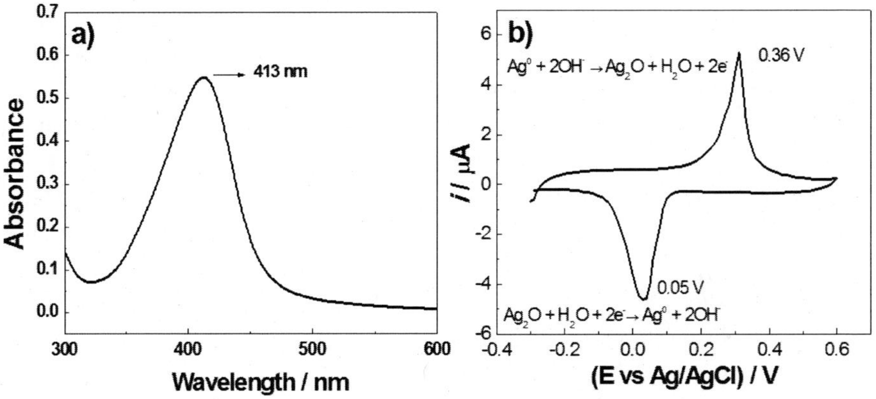

Figure 2. (a) UV-visible spectrum of Ag-NP. (b) CV of Ag-NP in KOH 0.1 mol L^{-1}, υ: 0.05 V s^{-1}.

Porous carbons materials are porous solids, which are classified according to the pore diameter size: pore diameters with sizes smaller than 2 nm are called micropores, those in the range from 2 nm to 50 nm are called mesopores and the higher ones at 50 nm they are macropores (Davis, 2002). Porous carbonaceous materials have been used in many areas of modern technology, including supercapacitors (Lee et al., 2005), adsorbents (Yang et al., 2016), gas storage (Lawes et al., 2015), fuel cells (Dai et al., 2015), sensors (Wilson & Islam, 2015) and as catalyst supports (Potphode et al., 2015). However, porous carbon materials present limitations, for example: the interconnection of pores is not adequate, decreasing the diffusion process on carbon surface (H.-J. Liu et al., 2011). This limitation is eliminated through the development of HPC, these carbons present a multimodal distribution of interconnected pores (macropores, mesopores and micropores), which facilitates ion transport due to the short distance between pores (You et al., 2014). Recently, HPCs exhibit diverse applications because macropores can serve as ion buffer reservoirs, providing a shorter diffusion distance, mesopores provide ion transport pathways with minimized resistance, and micropores enhance the electrical double layer (Xing et al., 2009; F. Xu et al., 2011; L. Yu et al., 2016). Until now, in the literature there are few examples of HPC as a sensor for the detection of free cyanide (A. A. Cárdenas-Riojas et al., 2019; Cárdenas Riojas et al., 2019; Hongyuan Zhang et al., 2020b).

In this study of Cardenas-Riojas, rigid templates of silicon oxide nanoparticles (SiO_2-NP) of 400 nm size, were used. The diameter of the Np-SiO_2 determined the pore size of the resulting carbon material. The SiO_2-NP were obtain by the Stôber method (Harris et al., 1990; W. Wang et al., 2003). In the initial step of HPC synthesis, SiO_2-NP were threated at 1000°C for 4 h to interconnect the nanoparticles, then an impregnation with the resorcinol/formaldehyde resin was made in a relation of 1 g of resorcinol per 1.6 mL of formaldehyde and 0.4 mL of sodium carbonate 0.1 mol L^{-1} as catalyst, this composite was treated at 100 °C to achieve the polymerization (Baena-Moncada et al., 2013); then it was carbonized at 900 °C under an inert atmosphere for 24 h. Finally, the Np-SiO_2 template was removed by treatment with 10% hydrofluoric acid in ethanol/water medium (1:1). The synthesized HPC using SiO_2-NP of 400 nm is named HPC_{400}. HPC_{400} was morphologically characterized by Scanning Electron Microscopy (SEM), obtaining the sizes and shapes of the pores.

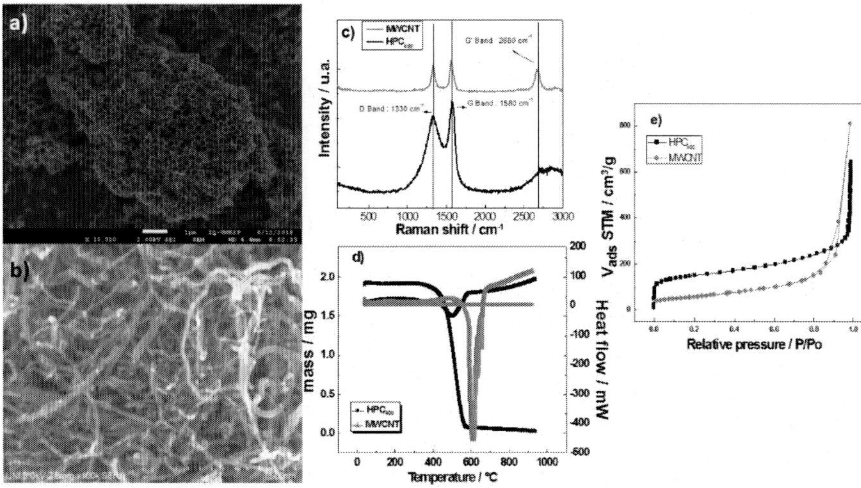

Figure 3. SEM images of (a) HPC_{400} and b) MWCNT; (c) Raman spectroscopy of HPC_{400} and MWCNT with a 532 nm laser and a power of 10%; (d) TGA-DSC analysis of HPC_{400} and MWCNT in medium oxygen in a temperature range 25 – 1000 °C. (e) BET Isotherms of HPC_{400} and MWCNT.

Figure 3a and b present the SEM images of the HPC_{400} and MWCNT. In addition, the interconnection of meso and macropores is appreciated, this characteristic of porous carbon is useful when used as a nanoparticles support and ions diffuse more easily. Figure 4c shows the Raman spectra of HPC_{400} and MWCNT, excited with a 532 nm laser. The intense band on 1580 cm^{-1}, corresponding to the fundamental vibration (first order) of tangential elongation called G and another weaker band at 2695 cm^{-1}, called G' corresponding to an overtone (second order), we also observe the presence of an intense band at 1330 cm^{-1}, called D ("Disorder induced"), and the band, D' + D, at 2900 cm^{-1}. The G and D first order bands indicate the sp^2 hybridization of HPC, in addition the weak second order bands G' and D' + D are due to the amorphous structure and the presence of asymmetry in the 3D network of the HPC (Dresselhaus et al., 2010).

Figure 3d presents the HPC_{400} and MWCNT thermal analyzes with differential scanning calorimetry (TGA-DSC) under oxygen atmosphere, in a temperature range of 25 – 1000 °C. In the temperature range below ~ 150°C, mass loss of HPC_{400} was observed by the TGA analysis, due to the humidity of the water molecules present in the environment and some volatile organic compounds, whereas the DSC analysis indicates an endothermic process due to decreased heat flow. At approximately ~ 150 to 400 °C, an endothermic process due to decreased heat flow occurs for the HPC_{400}. In the range of 400 – 600 °C, the total oxidation of the HPC occurred, and a notorious endothermic process was carried out with a large variation in heat flow due to oxidation. At temperatures higher than 600 °C, with less than ~ 5% mass of the HPC remaining, an exothermic process was carried out with an increase in heat flux and the formation of carbonaceous material to ash (Supan et al., 2017). In the study of MWCNT by TGA-DSC; Initially, a mass loss is observed, this occurs from room temperature to 200 °C, which is attributed to the removal of physisorbed water or CO_2. A second loss is observed up to approximately 500 °C, which can be attributed to the oxidation of amorphous carbon containing CHx species with defects in their structure. This loss is very small suggesting that the amount of carbon with disordered structures is very low. The total oxidation of the MWCNT occurs between 500 and 650 °C, generating a great variation in heat flow, due to the endothermic process. At temperatures higher than 650 °C, only carbon residue and ash formation remained.(Supan et al., 2017) The specific surface areas were calculated from the BET measurements of nitrogen adsorption isotherms (Brunauer-Emmett-Teller), obtaining specific surface areas of 535 m^2 g^{-1} and 204.9 m^2 g^{-1} for HPC_{400} and MWCNT, respectively, Figure 3e.

Figure 4 shows the CV of Ag-NP/HPC_{400} and Ag-NP/MWCNT, both figures present the characteristic oxidation potentials (0.31 V) and reduction (0.05 V) of Ag-NP (Toh et al., n.d.). A stability test was performed on these sensors to obtain the equilibrium time of the reaction between Ag-NP with the cyanide ion present in the different concentrations of free cyanide, 1×10^{-2} - 1×10^{-8} mol L^{-1} [CN^-]. The experiment was carried out in a two-electrode arrangement using Ag/AgCl as reference electrode. Figure 4c and 4d show that the potential decreases over time as the concentration of the cyanide ion increases, for both cases, 5 min was considered as a response in the equilibrium process. It was observed that the HPC_{400} sensor is more stable; due to the greater dispersion of the Ag-NP on the HPC porous surface.

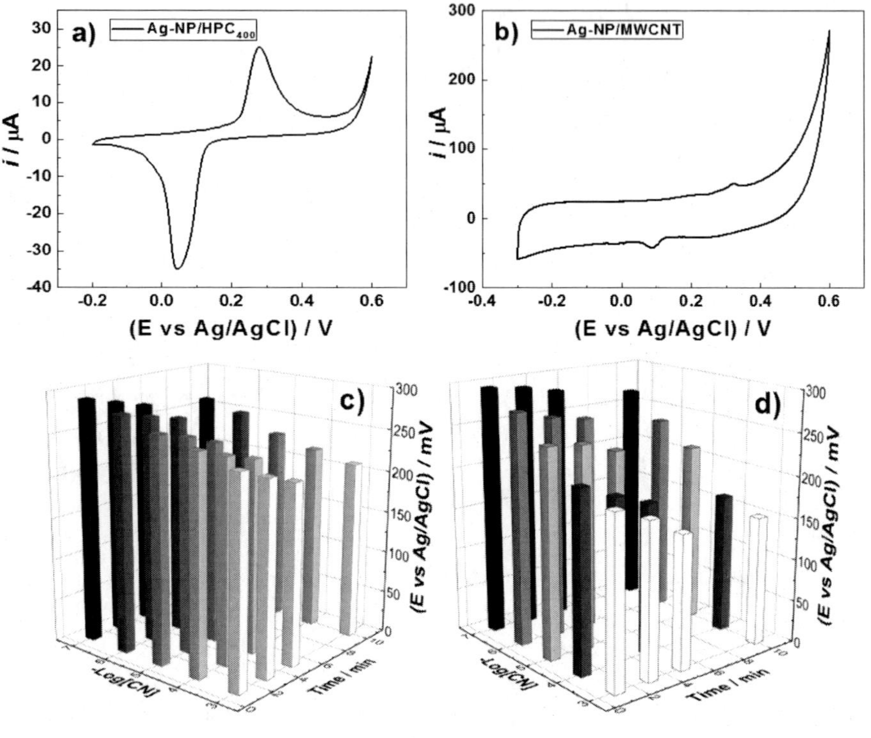

Figure 4. CV of (a) Ag-NP/HPC_{400}, and (b) Ag-NP/MWCNT using 0.1 mol L^{-1} KOH as electrolyte, at 20 mV s^{-1}. Stability of the different sensors: (c) Ag-NP/MWCNT and (d) Ag-NP/HPC_{400}, in a time interval in presence ion CN^-.

The reproducibility test of the sensors was performed in triplicate. The objective of this test is to obtain the same slope for each calibration curve, indicating that the sensor is reproducible, this experiment was carried out at different concentrations of free cyanide between 1×10^{-3} to 1×10^{-7} mol L^{-1}. The values obtained from the slopes of the calibration curves are 30.84 ± 0.96 mV and 16.18 ± 2.25 mV for Ag-NP/MWCNT and Ag-NP/HPC_{400} sensors, respectively (Figure 5), It is concluded that the Ag-NP/MWCNT sensor present a better response at each change in concentration, with respect to the Ag-NP/HPC_{400} sensor due to the greater interaction between Ag-NP and CN^-

. Additionally, both sensors present a linear working range between 1×10^{-3} to 1×10^{-7} mol L^{-1} of free cyanide ion $[CN^{-}]$ with the statistical r^2 close to 1, the results are indicated in Table 2 and Figure 5.

The detection limit and the quantification limit of both sensors Ag-NP/MWCNT and Ag-NP/HPC_{400} were calculated in the concentration range of 1×10^{-3} - 1×10^{-7} mol L^{-1}, using the method recommended by International Union of Pure and Applied Chemistry, IUPAC, (Buck & Lindner, 2009). The limit of detection (LOD) and the limit of quantification (LOQ) is 2.03 ± 0.22 ppb and 6.15 ± 0.68 ppb for Ag-NP/MWCNT, respectively. While Ag-NP/HPC_{400} presents and LOD of 1.75 ± 0.48 ppb and LOQ of 5.30 ± 1.46 ppb (A. A. Cárdenas-Riojas et al., 2019).

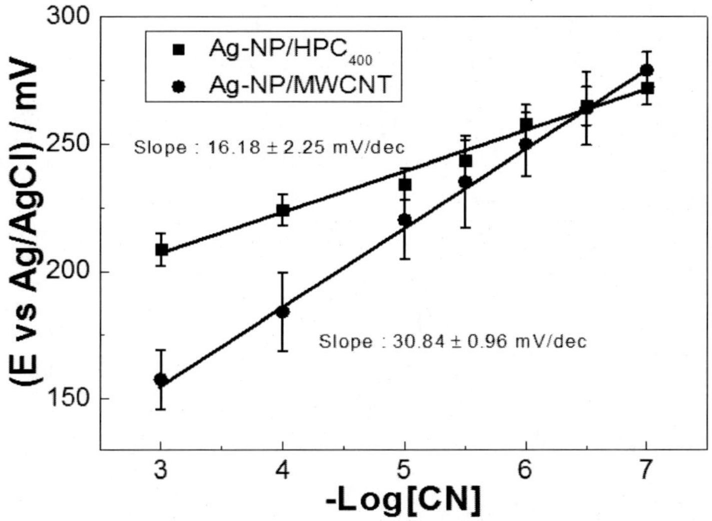

Figure 5. Results of Ag_2S-NP/HPC_{400} and Ag_2S-NP/MWCNT of cyanide concentrations 1×10^{-3} - 1×10^{-7} mol L^{-1} and linear, in electrolyte 0.1 mol L^{-1} KOH.

Table 2. Values obtained from the calibration curve of Ag-NP/MWCNT and Ag-NP/HPC_{400} sensors

Sensor	**Range Linear / mol L^{-1}**	**r^2**	**LOD / ppb**	**LOQ / ppb**
Ag-NP/MWCNT	1×10^{-3}-1×10^{-7}	0.9977	2.03 ± 0.22	6.15 ± 0.68
Ag-NP/HPC_{400}	1×10^{-3}-1×10^{-7}	0.9797	1.75 ± 0.48	5.30 ± 1.46

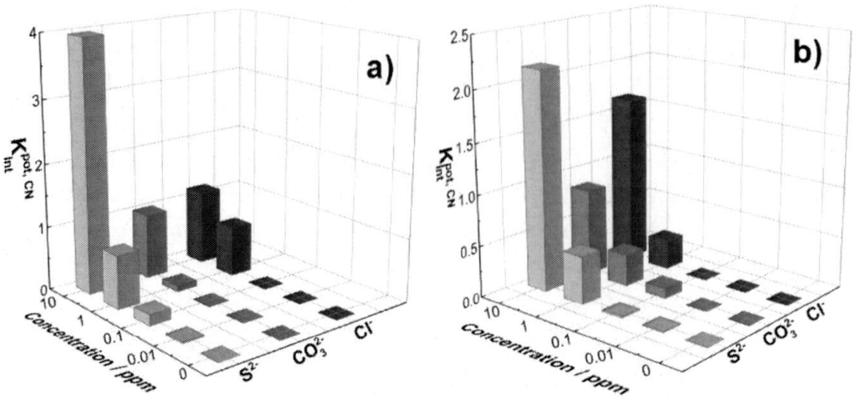

Figure 6. Results of the selectivity coefficients of the interferents Cl^-, $CO^{2-}{}_3$ and S^{2-} evaluated by the sensors (a) Ag-NP/HPC_{400} and (b) Ag-NP/MWCNT.

The Ag-NP/MWCNT and Ag-NP/HPC_{400} sensors, were tested with different ions as interferents, Cl^-, CO_3^{2-} and S^{2-} to obtain the degree of Selectivity of the sensors by the interferer ratio method, $K_{A,B}^{pot}$ was calculated to obtain the degree of selectivity from Nikolsky-Eisenman equation (Buck & Lindner, 2009). If $K_{A,B}^{pot} > 1$, the sensor responds selectively to the interferer rather than to the primary ion (CN^-); and if $K_{A,B}^{pot} < 1$, the sensor responds selectively to the primary ion (CN^-), rather than the interfering ion.

The degree of selectivity of Ag-NP/MWCNT and Ag-NP/HPC_{400} sensors with the different interferents are shown in Figure 6. The results obtained here, show that the sensors display high selectivity in the presence of various interferents. The Ag-NP/MWCNT sensor has selectivity constant values less than 1 for concentrations less than 10 ppm. The Ag-NP/HPC_{400} sensor presented selectivity constant values less than 1 for concentrations less than 1 ppm. From the tests carried out on both sensors, Ag-NP/MWCNT and Ag-NP/HPC_{400}, it was observed that each sensor presents particular advantages, for example the Ag-NP/HPC_{400} sensor displays better stability, detection limit, wheras the Ag-NP/MWCNT sensor exhibits a better linear range and better response with interferents.

3. Sensors Based on Silver Sulfide Nanoparticles Supported on Carbon Materials for the Potentiometric and Amperometric Detection of Free Cyanide

Silver sulfide nanoparticles (Ag_2S-NP) have been prepared by numerous methods including the use of inorganic precursors, which provide Ag^+ and S^{2-} ions (Sadovnikov et al., 2015). Typically, a $AgNO_3$ salt is used as the source of silver, while organic and inorganic sulfides provide S^{2-} ions, for example: Na_2S, $(NH_4)_2S$, H_2S and alkyl thiols (Feng Gao et al., 2002; Martínez-Castañón et al., 2005). Surfactants, on the other hand, are used to form micelles or microemulsions in chemical synthesis in order to control the size and shape of the nanoparticles (Armelao et al., 2002).

The chemistry of Ag_2S-NP presents greater thermodynamic stability than silver nanoparticles, in addition to a reduction in the dissolution of Ag^+, reaction 4 present the formation of silver ions, Ag^+, from silver sulfide (J. Liu et al., 2012; Ratte, 1999).

$$\mathbf{Ag_2S_{(s)} \rightarrow 2Ag^+ + S^{2-}, K_{ps} = 6 - 8 \times 10^{-51}} \qquad \text{reaction 4}$$

In this work, Ag_2S-NP are used for the development of sensors for the electrochemical detection of free cyanide, this process is due to the fact that Ag^+ ions form a complex with cyanide at basic pH, $[Ag(CN)_2]^-$. The complexation reaction 5 is:

$$\mathbf{Ag^+ + 2CN^- \leftrightharpoons [Ag(CN)_2]^- \ \beta = 2.8 \times 10^{20}} \qquad \text{reaction 5}$$

The synthesis of the Ag_2S-NP supported on different carbons materials was carried out by the coprecipitation method, 200 mL of $AgNO_3$ 8.0 mmol L^{-1} was prepared under constant stirring, 250 mg of carbon material was added; then, 20 mL of ammonium sulfide 4 mmol was added dropwise, at room temperature in an inert atmosphere of N_2. Additionally, unsupported Ag_2S-NP were synthetized in the same way but without carbon material (A. A. Cárdenas-Riojas et al., 2019; Martínez-Castañón et al., 2005). For this experiment, HPC_{400}, HPC_{300} and MWCNT were used as supporting materials. To obtain HPC_{300}, SiO_2-NP of 300 nm size was used as rigid template following the same procedure described in section 2 for HPC_{400}.

Figure 7 presents SEM images of the synthetized sensor, in the case of Ag_2S-NP/HPC_{300} sensor, it can be observed large aggregates of NP-Ag_2S

supported on HPC-300, with an approximate size of 120 - 150 nm (Figure 7a). In contrast to Ag_2S-NP/HPC_{400} sensor, Ag_2S nanowires are observed (Figure 7b), these nanowires were formed during the synthesis of Ag_2S-NP on HPC_{400}, using as template the mesopores presence on carbon structure. Due to the formation of these Ag_2S nanowires, fractures on HPC_{400} occur, as observed in their SEM images the diameter size of these Ag_2S nanowires is ~ 50 nm. The formation of Ag_2S nanowires was corroborated by studies conducted by (Du et al., 2007). Figures 7c presents the SEM images of Ag_2S-NP/MWCNT sensor, the Ag_2S-NP have a size of ~ 50 - 150 nm, preserving the morphology and size similar to that obtained with the unsupported Ag_2S-NP (Figure 7d).

In order to determine the oxidation and reduction states of silver in Ag_2S-NP supported on carbonaceous materials, an electrochemical characterization by CV was carried out (Figure 9). An oxidation peak from Ag^0 to Ag^{+1} at 0.31 V was observed in all cases; which is characteristic of silver (Cloake et al., 2015; Taheri et al., 2009). Figure 8 shows EDS for Ag_2S crystals confirming the presence of Ag and S in Ag_2S-NP.

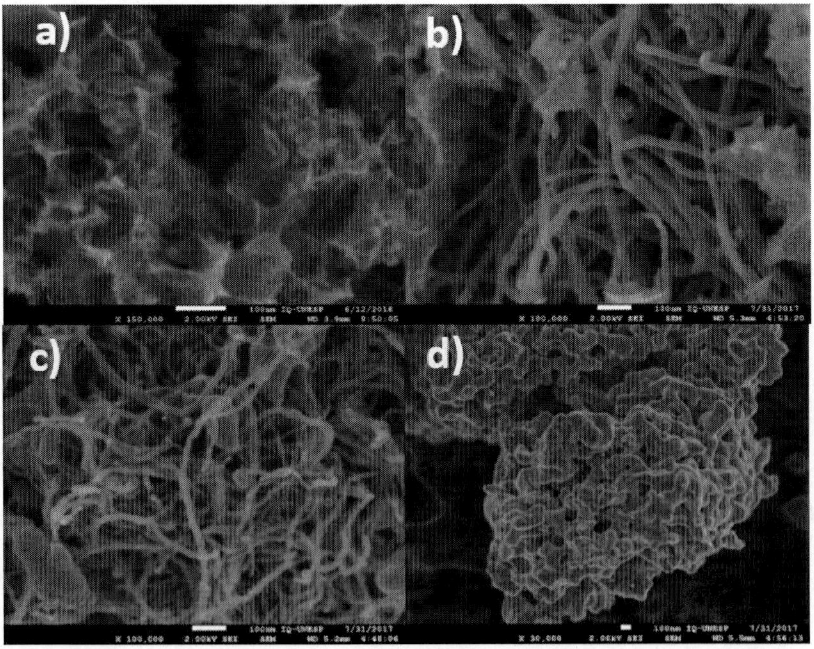

Figure 7. SEM images of (a) Ag_2S-NP/HPC_{300}, (b) Ag_2S-NP/HPC_{400}; (c) Ag_2S-NP/MWCNT and (d) Ag_2S-NP.

3.1. Potentiometric Detection of Free Cyanide

The potentiometric analysis was carried out on Ag_2S-NP/HPC_{400} and Ag_2S-NP/MWCNT sensors, since the HPC_{400} carbon has a greater electroactive and surface area with respect to the other carbons, this improves the diffusion of the ions on porous carbon surface, presenting a potential difference notorious when detecting free cyanide (Ma et al., 2016), to study the effect of different types of carbon materials morphologies in the electrochemical free cyanide detection, a comparison was made with the Ag_2S-NP/MWCNT sensor.

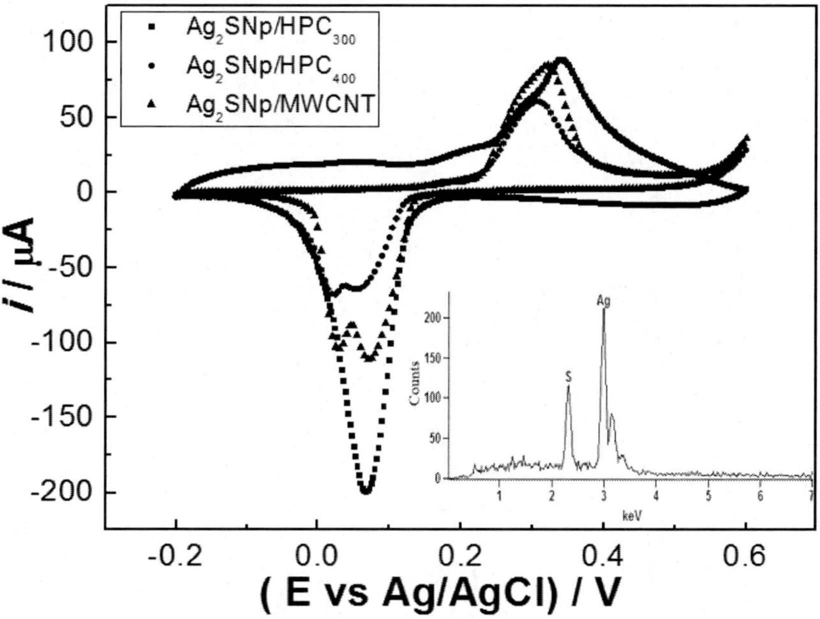

Figure 8. CV in 0.1 mol L^{-1} KOH at 0.02 V s^{-1} of the Ag_2S-NP/HPC_{300}, Ag_2S-NP/HPC_{400} and Ag_2S-NP/MWCNT sensors, insert figure EDS analysis of Ag_2S-NP.

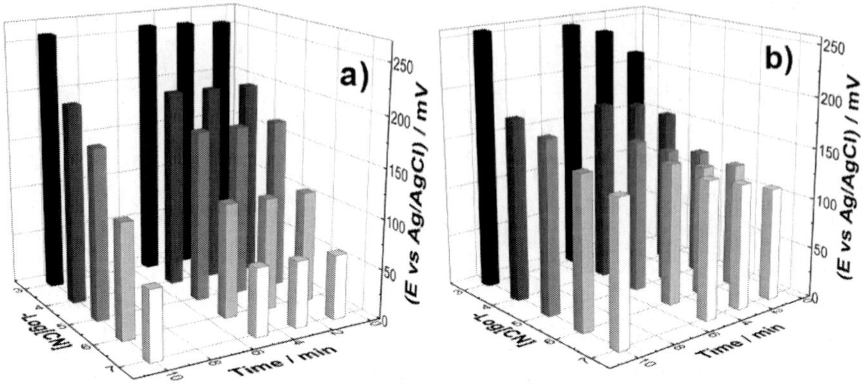

Figure 9. Stability tests of (a) Ag_2S-NP/HPC_{400} and (b) Ag_2S-NP/MWCNT sensors, condition range of cyanide concentrations $1\times10^{-3} - 1\times10^{-7}$ mol L^{-1}, 10 min of time application in KOH 0.1 mol L^{-1}.

A stability test was performed on Ag_2S-NP/HPC_{400} and Ag_2S-NP/MWCNT sensors to determine the necessary time to reach the reaction equilibrium between the Ag_2S-NP and CN^- ion at different concentrations of the free cyanide ion (1×10^{-8} - 1×10^{-2} mol L^{-1}), the results are shown in Figure 9. In Figure 9a it can be observed that for Ag_2S-NP/HPC_{400}, 1 min is necessary to reach the equilibrium in the presence of the cyanide ion. Additionally, a constant potential is observed all time and with the increase of CN^- concentration. In the case of Ag_2S-Np/MWCNT (Figure 9b), it is also observed that the potential increases with the relation of time and also with increasing cyanide concentration and reaches equilibrium in the presence of the cyanide ion after 5 min, indicating that the reaction equilibrium is slow. Probably due to a better dispersion of Ag_2S-NP on HPC porous surface; Ag_2S-NP/HPC_{400} is more stable than Ag_2S-NP/MWCNT.

The commercial ISE sensor presents a linear range of 1×10^{-7} - 1×10^{-3} mol L^{-1} [CN^-] with a slope of 53.1 mV/dec, and a linear correlation coefficient r^2 = 0.993 similar to the manual of the commercial electrode (A. Cárdenas-Riojas et al., 2019). This value was taken as a similar for the reproducibility of the developed sensors. The reproducibility test was performed in triplicate in order to obtain an average of the slopes of Ag_2S-NP/HPC_{400} and Ag_2S-NP/MWCNT. The slopes obtained were compared with the results of the ISE commercial electrode. The average value of the slopes is -22.30 ± 2.12 mV/dec and -43.82 ± 6.42 mV/dec for Ag_2S-NP/MWCNT and Ag_2S-NP/HPC_{400}, respectively. Figure 10 shows the reproducibility results of Ag_2S-NP/HPC_{400}, similar slope values were obtained for each test performed. Figure 10 presents the study of the linear range for Ag_2S-NP/HPC_{400}, obtaining a cyanide concentration range of 1×10^{-7} - 1×10^{-3} mol L^{-1} [CN^-], with a linear correlation coefficient value of $r^2 = 0.9909$. Its detection limit is 2.67 ± 0.17 μg L^{-1} and its detection quantification is 8.82 ± 0.56 μg L^{-1}; the study of the linear range for Ag_2S-NP/MWNCT sensor, obtaining a cyanide concentration range of 1×10^{-7} - 1×10^{-5} mol L^{-1} [CN^-], with a correlation coefficient value linear $r^2 = 0.9909$. Its detection limit is 5.02 ± 2.78 μg L^{-1} and its detection quantification is 16.83 ± 8.81μg L^{-1}, Table 3.

Table 3. Values obtained from the limits of detection and quantification of Ag_2S-NP/HPC_{400} and Ag_2S-NP/MWCNT sensors

Sensor	Range Linear / mol L^{-1}	r^2	LOD / μg L^{-1}	LOQ / μg L^{-1}
Ag_2S-NP/MWCNT	1×10^{-5}-1×10^{-7}	0.9798	5.02 ± 2.78	16.83 ± 8.81
Ag_2S-NP/HPC_{400}	1×10^{-3}-1×10^{-7}	0.9909	2.67 ± 0.17	8.82 ± 0.56

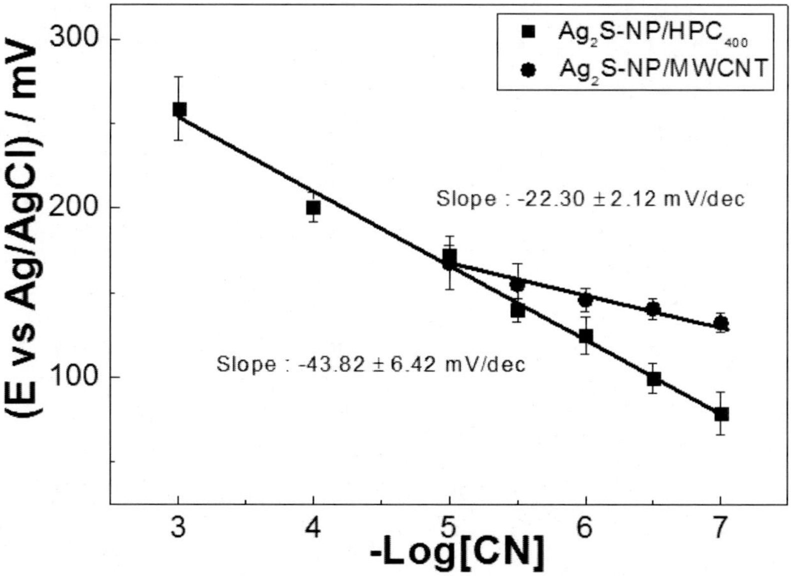

Figure 10. Results of Ag_2S-NP/HPC_{400} and Ag_2S-NP/MWCNT of cyanide concentrations 1×10^{-3} – 1×10^{-7} mol L^{-1} and linear, in electrolyte 0.1 mol L^{-1} KOH.

The selectivity degrees of Ag_2S-NP/HPC_{400} and Ag_2S-NP/MWCNT with the different interferents are shown in Figure 11, applying equation 1. (Buck & Lindner, 2009). The results here obtained, show that the sensors display high selectivity in the presence of various interferents. The Ag_2S-NP/MWCNT has selectivity constant values less than 1 for concentrations less than 1 ppm. The Ag_2S-NP/HPC_{400} sensor presented selectivity constant values less than 1 for concentrations less than 10 ppm.

The detection and quantification limits obtained in the previous reports are lower than those obtained in this work (Olazo-Quispe & La Rosa-Toro Gómez, 2014; Shamsipur et al., 2017; Taheri et al., 2009), due to the agglomeration of the Ag_2S-NP on MWCNT surface, also there is a decrease of its conductivity due to Ag_2S-NP presence. However, they are within the limits established by environmental quality standards in Peru of 22 μg L^{-1} (Ministerio del Medio Ambiente Perú & El Peruano, 2015).

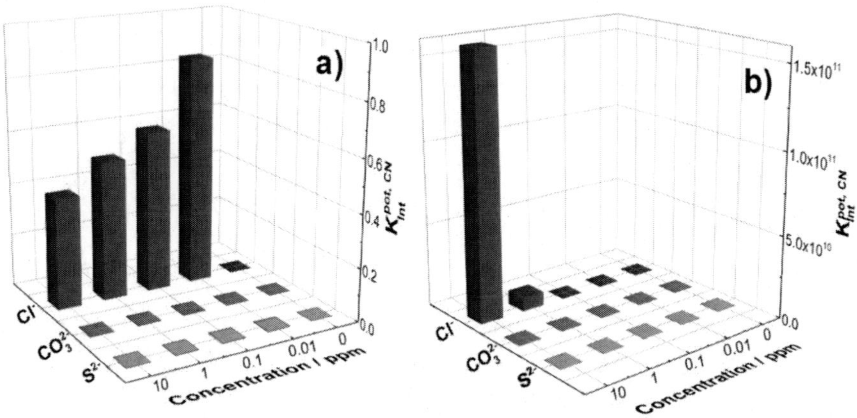

Figure 11. Results of the selectivity coefficients of the interferents Cl^-, CO^{2-}_3 and S^{2-} evaluated by the sensors a) Ag_2S-NP/HPC_{400} and b) Ag_2S-NP/MWCNT.

3.2. Amperometric Detection of Free Cyanide

The Ag_2S-NP/HPC_{300} was mixed with powdered graphite, G, to form the carbon paste for amperometric measurements. 50 mg of graphite powder and 50 mg of Ag_2S-NP/HPC_{300} were mixed in 500 μL of 0.05 mol L^{-1} KCl, until a homogeneous mixture was formed. Then it was dried at ~ 60 °C for 30 mins.

Then, 100 μL of Nujol oil was added to the mixture, it was mixed until achieving a homogeneous paste, obtained Ag_2S-NP/HPC_{300}/G sensor.

Different tests were carried out by CV in order to obtain the electrochemical behavior of Ag_2S-NP/HPC_{300}/G sensor in the presence of free cyanide. Figure 12 shows the cyclic voltammetry of the Ag_2S-NP/HPC_{300}/G in the absence and presence of cyanide in KOH 0.1 mol L^{-1} as supporting electrolyte at a scan rate of 0.05 V s^{-1}. In the absence of cyanide, a maximum anodic current peak was obtained at 0.38 V due to the oxidation reaction of Ag^0 to Ag^{+1}, and a maximum cathodic current at 0.8 V corresponding to the reduction reaction Ag^0 to Ag^{+1}. In the presence of the cyanide ion, a decrease in the maximum anodic and cathodic current occurs due to the formation of $[Ag(CN)_2]^-$complex (reaction 5), which is not electroactive. When it was added a higher concentration of free cyanide, a greater inhibition of current peaks was observed.

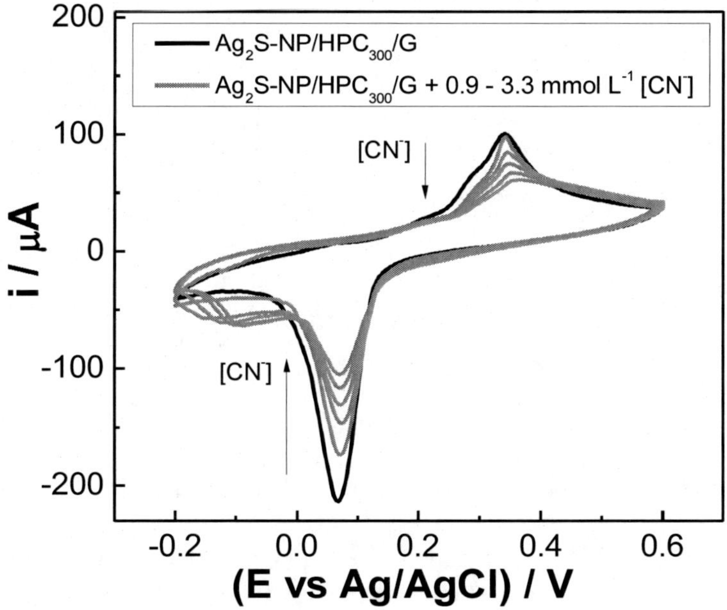

Figure 12. CV Ag_2S-NP/HPC_{300}/G sensor, in KOH 0.1 mol L^{-1} at 0.05 V s^{-1}, with different concentrations of ion [CN^-].

The stability of the Ag_2S/HPC_{300}/G sensor was obtained by SWV using the electrochemical signal of the oxidation peak (Ag^0 to Ag^{+1}), where the peak signal was verified by 40 measurements carried out in the absence of cyanide.

These results are presented in Table 4 where the values obtained from the relative standard deviations (RSD) for each of the tests performed are presented. These values indicate that the sensor has adequate stability.

The Ag_2S-NP/HPC_{300}/G sensor is reproducible after 10 measurements by SWV in the presence of different concentrations of cyanide (2×10^{-5} mol L^{-1} [CN^-] and b) 2×10^{-4} mol L^{-1} [CN^-]) as shown in Table 5 (Cárdenas Riojas et al., 2019).

The interaction of Ag_2S-NP/HPC_{300}/G sensor with the cyanide ion was evaluated. it is obtained a LOD of 7.3×10^{-8} mol L^{-1} (1.89 μg L^{-1}) and LOQ of 2.4×10^{-7} mol L^{-1} (6.24 μg L^{-1}) from the analysis of Figure 13 and Table 6. It also displays two linear ranges, 9.4×10^{-7} - 1.06×10^{-4} mol L^{-1} and the other of 1.06×10^{-4} - 6.51×10^{-4} mol L^{-1}.

Table 4. Ag_2S-NP/HPC300/G sensor stability results

	Test 1	**Test 2**	**Test 3**
Medium	424.04	392.14	377.70
Standard deviation	16.39	20.69	15.74
Relative Standard Deviation	0.03	0.05	0.04
RSD×100%	3.87	5.28	4.17

Table 5. Results of the stability of Ag_2S-NP/HPC_{300}/G sensor in the presence of cyanide

	2×10^{-5} mol L^{-1} [CN^-]	**2×10^{-4} mol L^{-1} [CN^-]**
Medium	217.98	47.63
Standard deviation	7.92	5.00
Relative Standard Deviation	0.03	0.10
RSD × 100%	3.63	10.51

Real samples tests were carried out with water from rivers in Peru and Brazil, these waters were enriched with three different concentrations of cyanide. The samples were added to the electrochemical cell and analyzed directly using the SWV method. Furthermore, the samples were diluted 50 times in these experiments. Figure 14 presents the results obtained for the Ag_2S-NP/HPC_{300}/G sensor and the cyanide concentrations obtained with the real samples, these concentrations indicate that the sensor is suitable for studies with real samples. In addition these experiments were performed in triplicate.

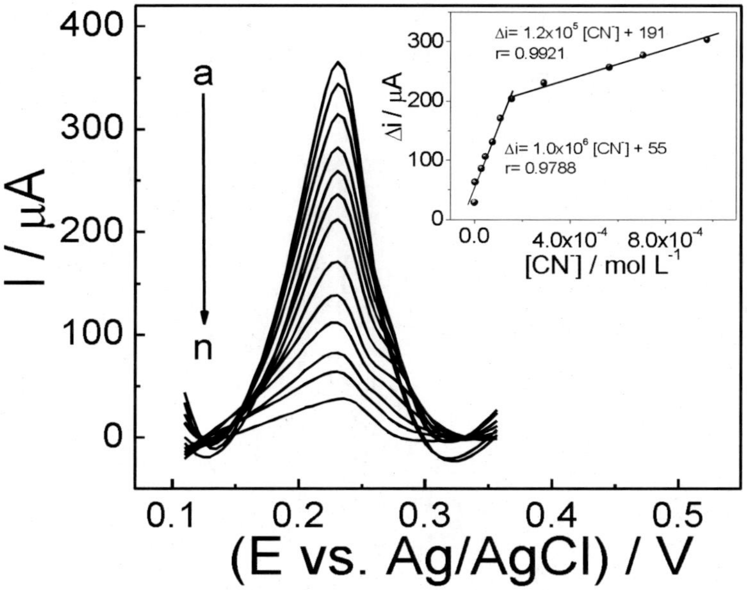

Figure 13. Ag_2S-NP/HPC_{300}/G sensor calibration curve. This Figure was reproduced with permission from Ref (Cárdenas Riojas et al., 2019).

Table 6. Values of the detection and quantification limits of Ag_2S-NP/HPC_{300}/G sensor

Curve	Linear Range 1	Linear Range 2	LOD	LOQ
	(mol L^{-1})		(mol L^{-1})	
1	9.8×10^{-7} – 1.53×10^{-4}	1.53×10^{-4} - 9.74×10^{-4}	4.0×10^{-8}	1.3×10^{-7}
2	9.1×10^{-7} – 1.37×10^{-4}	1.37×10^{-4} - 8.89×10^{-4}	9.7×10^{-8}	3.2×10^{-7}
3	9.1×10^{-7} – 2.94×10^{-5}	2.94×10^{-5} - 1.00×10^{-4}	8.6×10^{-8}	2.8×10^{-7}
Average	9.4×10^{-7} – 1.06×10^{-4}	1.06×10^{-4}- 6.51×10^{-4}	7.3×10^{-8}	2.4×10^{-7}

The amperometric sensor made as carbon paste, Ag_2S-NP/HPC_{300}/G, is compared with other sensors reported in the literature for the detection of free cyanide. The detection and quantification limits obtained in different works reported in the literature are in the range of 1.4×10^{-8} to 9.0×10^{-8} mol L^{-1} and are comparable with this work (7.3×10^{-8}) (Ali et al., 2017; Shamsipur et al.,

2017; Taheri et al., 2009; S. Wang et al., 2010) being within the limits established by environmental quality standards in Peru of 22 μg L^{-1}.

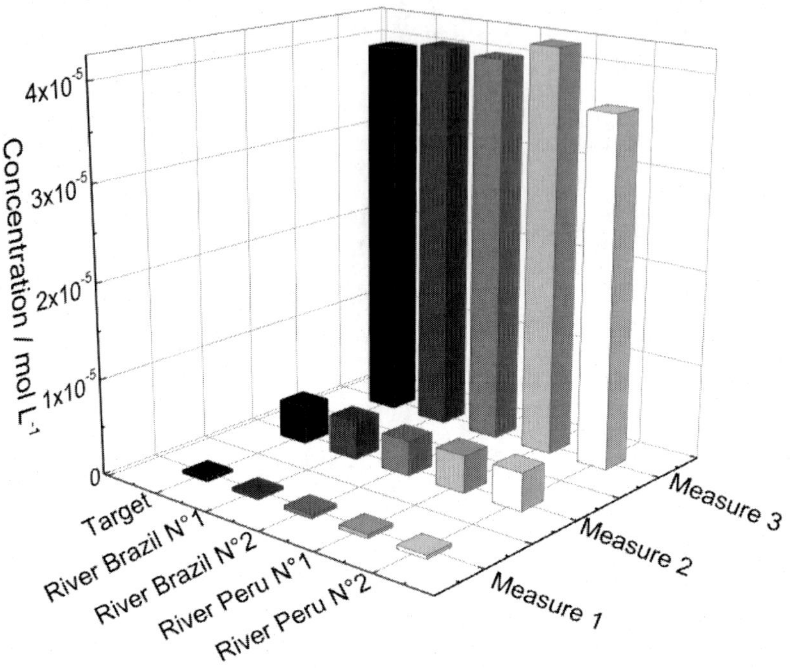

Figure 14. Analysis of real samples by SWV of the rivers of Brazil, and the Tingo-Maygasbamba river of Cajamarca – Peru.

Conclusion

Potentiometric sensors were developed using the layer-by-layer self-assembly method, using PDADMAC to anchor Ag-NP on the carbonaceous materials HPC_{400} and MWCNT surface. The Ag-NP/HPC_{400} and Ag-NP/MWCNT sensors were characterized by different electrochemical and physicochemical techniques. The limit of detection and the limit of quantification were 2.03 ± 0.22 ppb and 6.15 ± 0.68 ppb for Ag-NP/MWCNT, respectively. While Ag-NP/HPC_{400} presents and LOD of 1.75 ± 0.48 ppb and LOQ of 5.30 ± 1.46 ppb.

Also, this work successfully describes the development of a new sensor platform for the sensitive detection of cyanide in alkaline medium based on a hierarchical porous carbon modified with Ag_2S-NP. The proposed analytical method is proved to be highly effective, practical and precise in the determination of free cyanide, with excellent sensitivity and selectivity, mainly when applied to river water samples.

The Ag_2S-NP/HPC_{400} sensor was evaluated as a potentiometric sensor. This sensor has a LOD = 3.34 μg L^{-1} and LOQ = 11.05 μg L^{-1} were obtained, in a linear working range of 1×10^{-7} - 1×10^{-3} mol L^{-1}. Likewise, the Ag_2S-NP/MWCNT sensor presented a LOD = 9.22 μg L^{-1} and LOQ = 30.42 μg L^{-1}, in a linear working range of 1×10^{-7} - 1×10^{-5} mol L^{-1}, presenting a lower linear range than the sensor supported in the nested porous carbon HPC_{400}, as well as a higher detection limit.

The Ag_2S-NP/HPC_{300}/G sensor was studied as an amperometric sensor. This sensor has a LOD = 1.89 μg L^{-1} and LOQ = 6.27 μg L^{-1}. It also presents two linear ranges, 9.4×10^{-7} - 1.06×10^{-4} mol L^{-1} and 1.06×10^{-4} - 6.51×10^{-4} mol L^{-1}.

When comparing the three carbonaceous materials, the MWCNT, HPC_{300} and HPC_{400}, it can be clearly seen that there is an effect of the support on the performance of the sensor. These differences may be attribued to the characteristics of the support, where the hierarchical porous carbon allows a better diffusion of the electrolyte and better access to the active sites, in addition the HPC present a much more developed area than the MWCNT, as well as a better adherence between the Ag_2S-NP and the HPC support. Furthermore, the proposed method can determine cyanide in environmental samples for both potenciometric and amperometric detection without the need for prior treatment. This method is faster than many other analytical procedures reported in the literature.

References

Afkhami, A. & Sarlak, N. (2007). A novel cyanide sensing phase based on immobilization of methyl violet on a triacetylcellulose membrane. *Sensors and Actuators, B: Chemical*, *122*(2), 437–441. https://doi.org/10.1016/j.snb.2006.06.012.

Agarwalla, H., Gangopadhyay, M., Sharma, D. K., Basu, S. K., Jadhav, S., Chowdhury, A. & Das, A. (2015). Fluorescent probes for the detection of cyanide ions in aqueous medium: Cellular uptake and assay for β-glucosidase and hydroxynitrile lyase. *Journal of Materials Chemistry B*, *3*(47), 9148–9156. https://doi.org/10.1039/c5tb01853f.

Alexeyeva, N. & Tammeveski, K. (2008). Electroreduction of oxygen on gold nanoparticle/PDDA-MWCNT nanocomposites in acid solution. *Analytica Chimica Acta*, *618*(2), 140–146. https://doi.org/10.1016/J.ACA.2008.04.056.

Ali, T. A., Mohamed, G. G., Saber, A. L. & Almazroai, L. S. (2017). ELECTROCHEMICAL SCIENCE Potentiometric Determination of Cyanide in Polluted Water Samples Using Screen-Printed Electrode modified with Ruthenium(II) Complexes Ionophores. *International Journal of Int. J. Electrochem. Sci*, *12*, 11904–11919. https://doi.org/10.20964/2017.12.26.

Alizadeh, A., Ghouzivand, S., Khodaei, M. M. & Ardalani, M. (2016). An interesting spectroscopic method for chromofluorogenic detection of cyanide ion in aqueous solution: Disruption of intramolecular charge transfer (ICT). *Journal of Chemical Sciences*, *128*(4), 537–543. https://doi.org/10.1007/s12039-016-1051-y.

Alvillo-Rivera, A., Garrido-Hoyos, S., Buitrón, G., Thangarasu-Sarasvathi, P. & Rosano-Ortega, G. (2021). Biological treatment for the degradation of cyanide: A review. *Journal of Materials Research and Technology*, *12*, 1418–1433. https://doi.org/10.1016/j.jmrt.2021.03.030.

Armelao, L., Bertoncello, R., Cattaruzza, E., Gialanella, S., Gross, S., Mattei, G., Mazzoldi, P. & Tondello, E. (2002). Chemical and physical routes for composite materials synthesis: Ag and Ag_2S nanoparticles in silica glass by sol–gel and ion implantation techniques. *J. Mater. Chem.*, *12*(8), 2401–2407. https://doi.org/10.1039/B203539C.

Baena-Moncada, A. M., Morales, G. M., Barbero, C., Planes, G. A., Florez-Montaño, J. & Pastor, E. (2013). Formic acid oxidation over hierarchical porous carbon containing PtPd catalysts. *Catalysts*, *3*(4). https://doi.org/10.3390/catal3040902.

Bhaskar, R. & Sarveswari, S. (2019). Colorimetric sensor for real-time detection of cyanide ion in water and food samples. *Inorganic Chemistry Communications*, *102*(February), 83–89. https://doi.org/10.1016/j.inoche.2019.02.002.

Bläsing, K., Harloff, J., Schulz, A., Stoffers, A., Stoer, P. & Villinger, A. (2020). Salts of HCN-Cyanide Aggregates: [CN(HCN)2]− and [CN(HCN)3]−. *Angewandte Chemie - International Edition*, *59*(26), 10508–10513. https://doi.org/10.1002/anie.201915206.

Bolarinwa, I. F., Oke, M. O., Olaniyan, S. A. & Ajala, A. S. (2016). A Review of Cyanogenic Glycosides in Edible Plants. *Toxicology - New Aspects to This Scientific Conundrum*. https://doi.org/10.5772/64886.

Breuer, P. L., Sutcliffe, C. A. & Meakin, R. L. (2011). Cyanide measurement by silver nitrate titration: Comparison of rhodanine and potentiometric end-points. *Hydrometallurgy*, *106*(3–4), 135–140. https://doi.org/10.1016/j.hydromet.2010.12.008.

Buck, R. P. & Lindner, E. (2009). Recommendations for nomenclature of ionselective electrodes (IUPAC Recommendations 1994). *Pure and Applied Chemistry*, *66*(12), 2527–2536. https://doi.org/10.1351/pac199466122527.

Cárdenas-Riojas, A. A., Wong, A., Sotomayor, M. D. P. T., La Rosa-Toroa, A. & Baena-Moncada, A. M. (2019). Potentiometric sensor based on silver sulfide nanoparticles supported in carbonaceous material for the detection of free cyanide. *Quimica Nova*, *42*(3), 255–261. https://doi.org/10.21577/0100-4042.20170338.

Cárdenas-Riojas, A., Wong, A., Sotomayor, M. D. P., Rosa-Toro, A. & Baena-Moncada, A. (2019). Sensor Potenciométrico Basado En Nanopartículas De Sulfuro De Plata Soportadas En Materiales Carbonosos Para La Detección De Cianuro Libre. *Química Nova*. https://doi.org/10.21577/0100-4042.20170338. [Potentiometric Sensor Based on Silver Sulfide Nanoparticles Supported on Carbonaceous Materials for the Detection of Free Cyanide. *New Chemistry*.]

Cárdenas Riojas, A. A., Wong, A., Planes, G. A., Sotomayor, M. D. P. T., La Rosa-Toro, A. & Baena-Moncada, A. M. (2019). Development of a new electrochemical sensor based on silver sulfide nanoparticles and hierarchical porous carbon modified carbon paste electrode for determination of cyanide in river water samples. *Sensors and Actuators B: Chemical*, *287*, 544–550. https://doi.org/10.1016/J.SNB.2019.02.053.

Carter, S. R. & Rimmer, S. (2004). Surface molecularly imprinted polymer core-shell particles. *Advanced Functional Materials*, *14*(6), 553–561. https://doi.org/10.1002/adfm.200305069.

Chinaka, S., Tanaka, S., Takayama, N., Tsuji, N., Takou, S. & Ueda, K. (2001). High-Sensitivity Analysis of Cyanide by Capillary Electrophoresis with Fluorescence Detection. *Analytical Sciences*, *17*(5), 649–652. https://doi.org/10.2116/analsci.17.649.

Cloake, S. J., Toh, H. S., Lee, P. T., Salter, C., Johnston, C. & Compton, R. G. (2015). Anodic Stripping Voltammetry of Silver Nanoparticles: Aggregation Leads to Incomplete Stripping. *ChemistryOpen*, *4*(1), 22–26. https://doi.org/10.1002/open.201402050.

Cruz, D. A., Rodríguez, M. C., López, J. M., Herrera, V. M., Orive, A. G. & Creus, A. H. (n.d.). *Nanopartículas Metálicas Y Plasmones De Superficie: Una Relación Profunda Metallic Nanoparticles And Surface Plasmons: A Deep Relationship*. Retrieved July 17, 2021, from http://www.exeedu.com/publishing.cl/av_cienc_ing/67.

Dai, X., Chen, D., Fan, H., Zhong, Y., Chang, L., Shao, H., Wang, J., Zhang, J. & Cao, C. (2015). Ni(OH)2/NiO/Ni composite nanotube arrays for high-performance supercapacitors. *Electrochimica Acta*, *154*(Supplement C), 128–135. https://doi.org/10.1016/j.electacta.2014.12.066.

Dash, R. R., Gaur, A. & Balomajumder, C. (2009). Cyanide in industrial wastewaters and its removal: A review on biotreatment. *Journal of Hazardous Materials*, *163*(1), 1–11. https://doi.org/10.1016/j.jhazmat.2008.06.051.

Davis, M. E. (2002). Ordered porous materials for emerging applications. *Nature*, *417*(6891), 813–821. https://doi.org/10.1038/nature00785.

Decher, G. & Hong, J. -D. (1991). Buildup of ultrathin multilayer films by a self-assembly process, 1 consecutive adsorption of anionic and cationic bipolar amphiphiles on charged surfaces. *Makromolekulare Chemie. Macromolecular Symposia*, *46*(1), 321–327. https://doi.org/10.1002/MASY.19910460145.

Dehnicke, K. (1976). The Chemistry of Cyano Complexes of the Transition Metals. Organometallic Chemistry-A Series of Monographs. Von AG Sharpe. Academic Press, London-New York-San Francisco 1976. 1. Aufl., XI, 302 S., geb.£ 10.40. *Angewandte Chemie*, *88*(22), 774.

Dhahi, T. H. S., Bin Hashim, U. D. A., Ahmed, N. M. & Mat Taib, A. (2010). A review on the electrochemical sensors and biosensors composed of nanogaps as sensing material. *Journal of Optoelectronics and Advanced Materials*, *12*(9), 1857–1862.

Dresselhaus, M. S., Jorio, A., Hofmann, M., Dresselhaus, G. & Saito, R. (2010). Perspectives on carbon nanotubes and graphene Raman spectroscopy. In *Nano Letters* (Vol. *10*, Issue 3, pp. 751–758). American Chemical Society. https://doi.org/10.1021/nl904286r.

Du, N., Zhang, H., Sun, H. & Yang, D. (2007). Sonochemical synthesis of amorphous long silver sulfide nanowires. *Materials Letters*, *61*(1), 235–238. https://doi.org/10.1016/J.MATLET.2006.04.039.

Egekeze, J. O. & Oehme, F. W. (1980). Cyanides and their toxicity: a literature review. *Tijdschrift Voor Diergeneeskunde*, *105*(8), 37–41. https://doi.org/10.1080/01652176.1980.9693766.

Erdemir, S. & Malkondu, S. (2020). On-site and low-cost detection of cyanide by simple colorimetric and fluorogenic sensors: Smartphone and test strip applications. *Talanta*, *207*(May 2019), 120278. https://doi.org/10.1016/j.talanta.2019.120278.

Feng Gao,†, Qingyi Lu,† and & Zhao*, D. (2002). *Controllable Assembly of Ordered Semiconductor Ag2S Nanostructures*. https://doi.org/10.1021/NL025811A.

Figueira, E. C., Neres, L. C. S., Ruy, M. R. S., Troiano, G. F. & Sotomayor, M. D. P. T. (2016). Development of a biomimetic sensor for selective identification of cyanide. *Analytical Methods*, *8*(33), 6353–6360. https://doi.org/10.1039/c6ay01830k.

Ghanavati, M., Azad, R. R. & Mousavi, S. A. (2014). Amperometric inhibition biosensor for the determination of cyanide. *Sensors and Actuators, B: Chemical*, *190*, 858–864. https://doi.org/10.1016/j.snb.2013.09.055.

Gopal Reddy, C. V., Yan, F., Zhang, Y. & Vo-Dinh, T. (2010). A highly sensitive Raman method for selective cyanide detection based on evaporated cuprous iodide substrate. *Analytical Methods*, *2*(5), 458–460. https://doi.org/10.1039/c0ay00085j.

Hamel, J. (2011). A review of acute cyanide poisoning with a treatment update. *Critical Care Nurse*, *31*(1), 72–82. https://doi.org/10.4037/ccn2011799.

Harris, M. T., Brunson, R. R. & Byers, C. H. (1990). The base-catalyzed hydrolysis and condensation reactions of dilute and concentrated TEOS solutions. *Journal of Non-Crystalline Solids*, *121*(1), 397–403. https://doi.org/10.1016/0022-3093(90)90165-I.

Hernández, Y., Coello, Y., Fratila, R. M., de la Fuente, J. M. & Lionberger, T. A. (2017). Highly sensitive ratiometric quantification of cyanide in water with gold nanoparticles via Resonance Rayleigh Scattering. *Talanta*, *167*(February), 51–58. https://doi.org/10.1016/j.talanta.2017.02.006.

HISKEY, J. B. & ATLURI, V. P. (1988). Dissolution Chemistry of Gold and Silver in Different Lixiviants. *Mineral Processing and Extractive Metallurgy Review*, *4*(1–2), 95–134. https://doi.org/10.1080/08827508808952634.

Jaszczak, E., Narkowicz, S., Namieśnik, J. & Polkowska, Ż. (2017). Determination of cyanide in urine and saliva samples by ion chromatography with pulsed amperometric detection. *Monatshefte Fur Chemie*, *148*(9), 1645–1649. https://doi.org/10.1007/s00706-017-1977-x.

Jaszczak, E., Polkowska, Ż., Narkowicz, S. & Namieśnik, J. (2017). Cyanides in the environment—analysis—problems and challenges. *Environmental Science and Pollution Research*, *24*(19), 15929–15948. https://doi.org/10.1007/s11356-017-9081-7.

Jones, S. E. W., Campbell, F. W., Baron, R., Xiao, L. & Compton, R. G. (2008). Particle Size and Surface Coverage Effects in the Stripping Voltammetry of Silver Nanoparticles: Theory and Experiment. *Journal of Physical Chemistry C*, *112*(46), 17820–17827. https://doi.org/10.1021/JP807093Q.

Koch, W. F. (1983). Determination of Trace Levels of Cyanide By Ion Chromatography With Electrochemical Detection. *Journal of Research of the National Bureau of Standards (United States)*, *88*(3), 157–161. https://doi.org/10.6028/jres.088.008.

Lawes, S., Riese, A., Sun, Q., Cheng, N. & Sun, X. (2015). Printing nanostructured carbon for energy storage and conversion applications. *Carbon*, *92*(Supplement C), 150–176. https://doi.org/10.1016/j.carbon.2015.04.008.

Lawson-Smith, P., Jansen, E. C. & Hyldegaard, O. (2011). Cyanide intoxication as part of smoke inhalation - a review on diagnosis and treatment from the emergency perspective. *Scandinavian Journal of Trauma, Resuscitation and Emergency Medicine*, *19*, 1–5. https://doi.org/10.1186/1757-7241-19-14.

Leavesley, H. B., Li, L., Prabhakaran, K., Borowitz, J. L. & Isom, G. E. (2008). Interaction of cyanide and nitric oxide with cytochrome c oxidase: Implications for acute cyanide toxicity. *Toxicological Sciences*, *101*(1), 101–111. https://doi.org/10.1093/toxsci/kfm254.

Lee, J. W., Weiner, R. S., Sailstad, J. M., Bowsher, R. R., Knuth, D. W., O'Brien, P. J., Fourcroy, J. L., Dixit, R., Pandite, L., Pietrusko, R. G., Soares, H. D., Quarmby, V., Vesterqvist, O. L., Potter, D. M., Witliff, J. L., Fritche, H. A., O'Leary, T., Perlee, L., Kadam, S. & Wagner, J. A. (2005). Method Validation and Measurement of Biomarkers in Nonclinical and Clinical Samples in Drug Development: A Conference Report. *Pharmaceutical Research*, *22*(4), 499–511. https://doi.org/10.1007/s11095-005-2495-9.

Lindsay, A. E. & O'Hare, D. (2006a). The development of an electrochemical sensor for the determination of cyanide in physiological solutions. *Analytica Chimica Acta*, *558*(1–2), 158–163. https://doi.org/10.1016/j.aca.2005.11.036.

Lindsay, A. E. & O'Hare, D. (2006b). The development of an electrochemical sensor for the determination of cyanide in physiological solutions. *Analytica Chimica Acta*, *558*(1–2), 158–163. https://doi.org/10.1016/J.ACA.2005.11.036.

Liu, H.-J., Wang, J., Wang, C.-X. & Xia, Y.-Y. (2011). Ordered Hierarchical Mesoporous/Microporous Carbon Derived from Mesoporous Titanium-Carbide/Carbon Composites and its Electrochemical Performance in Supercapacitor. *Advanced Energy Materials*, *1*(6), 1101–1108. https://doi.org/10.1002/aenm.201100255.

Liu, J., Wang, Z., Liu, F. D., Kane, A. B. & Hurt, R. H. (2012). Chemical Transformations of Nanosilver in Biological Environments. *ACS Nano*, *6*(11), 9887–9899. https://doi.org/10.1021/nn303449n.

Logue, B. A., Hinkens, D. M., Baskin, S. I. & Rockwood, G. A. (2010). The analysis of cyanide and its breakdown products in biological samples. *Critical Reviews in Analytical Chemistry*, *40*(2), 122–147. https://doi.org/10.1080/10408340903535315.

Long, L., Yuan, X., Cao, S., Han, Y., Liu, W., Chen, Q., Han, Z. & Wang, K. (2019). Determination of Cyanide in Water and Food Samples Using an Efficient Naphthalene-Based Ratiometric Fluorescent Probe [Research-article]. *ACS Omega*, *4*(6), 10784–10790. https://doi.org/10.1021/acsomega.9b01308.

Lovasoa, C. R., Hela, K., Harinaivo, A. A. & Hamma, Y. (2017). Bioremediation of soil and water polluted by cyanide: A review. *African Journal of*

Environmental Science and Technology, *11*(6), 272–291. https://doi.org/10.5897/ajest2016.2264.

Lu, J., Qin, W., Zhang, Z., Feng, M. & Wang, Y. (1995). *A flow-injection type chemiluminescence-based cyanide sensor for*. *304*, 369–373.

Luque-Almagro, V. M., Blasco, R., Martínez-Luque, M., Moreno-Vivián, C., Castillo, F. & Roldán, M. D. (2011). Bacterial cyanide degradation is under review: Pseudomonas pseudoalcaligenes CECT5344, a case of an alkaliphilic cyanotroph. *Biochemical Society Transactions*, *39*(1), 269–274. https://doi.org/10.1042/BST0390269.

Lv, J., Zhang, Z., Li, J. & Luo, L. (2005). A micro-chemiluminescence determination of cyanide in whole blood. *Forensic Science International*, *148*(1), 15–19. https://doi.org/10.1016/j.forsciint.2004.03.032.

Ma, Y., Yu, B., Guo, Y. & Wang, C. (2016). Facile synthesis of biomass-derived hierarchical porous carbon microbeads for supercapacitors. *Journal of Solid State Electrochemistry*, *20*(8), 2231–2240. https://doi.org/10.1007/s10008-016-3233-4.

Manickam, S. & Iyer, S. K. (2020). Highly sensitive turn-off fluorescent detection of cyanide in aqueous medium using dicyanovinyl-substituted phenanthridine fluorophore. *RSC Advances*, *10*(20), 11791–11799. https://doi.org/10.1039/d0ra00623h.

Manoj, K. M., Ramasamy, S., Parashar, A., Gideon, D. A., Soman, V., Jacob, V. D. & Pakshirajan, K. (2020). Acute toxicity of cyanide in aerobic respiration: Theoretical and experimental support for murburn explanation. *Biomolecular Concepts*, *11*(1), 32–56. https://doi.org/10.1515/bmc-2020-0004.

Martínez-Castañón, G. A., Sánchez-Loredo, M. G., Dorantes, H. J., Martínez-Mendoza, J. R., Ortega-Zarzosa, G. & Ruiz, F. (2005). Characterization of silver sulfide nanoparticles synthesized by a simple precipitation method. *Materials Letters*, *59*(4), 529–534. https://doi.org/10.1016/j.matlet.2004.10.043.

Ministerio del Medio Ambiente Perú & El Peruano. (2015). *Decreto Supremo N°015-2015-MINAM*.

Nandi, L. G., Nicoleti, C. R., Marini, V. G., Bellettini, I. C., Valandro, S. R., Cavalheiro, C. C. S. & Machado, V. G. (2017). Optical devices for the detection of cyanide in water based on ethyl(hydroxyethyl)cellulose functionalized with perichromic dyes. *Carbohydrate Polymers*, *157*, 1548–1556. https://doi.org/10.1016/j.carbpol.2016.11.039.

Olazo-Quispe, R. & La Rosa-Toro Gómez, A. (2014). Micropartículas de Ag/Ag2S tipo core-shell como sensor potenciométrico para la detección de cianuro. *Revista de La Sociedad Química Del Perú*, *80*(1), 51–64.

Ou, X. X., Jin, Y. L., Chen, X. Q., Gong, C. Bin, Ma, X. B., Wang, Y. S., Chow, C. F. & Tang, Q. (2015). Colorimetric test paper for cyanide ion determination in real-time. *Analytical Methods*, *7*(12), 5239–5244. https://doi.org/10.1039/c5ay01033k.

Parker-Cote, J. L., Rizer, J., Vakkalanka, J. P., Rege, S. V. & Holstege, C. P. (2018). Challenges in the diagnosis of acute cyanide poisoning. *Clinical Toxicology*, *56*(7), 609–617. https://doi.org/10.1080/15563650.2018.1435886.

Podsiadlo, P. (2008). *Layer-by-layer assembly of nanostructured composites: Mechanics and applications*. University of Michigan.

Potphode, D. D., Sivaraman, P., Mishra, S. P. & Patri, M. (2015). Polyaniline/partially exfoliated multi-walled carbon nanotubes based nanocomposites for supercapacitors. *Electrochimica Acta*, *155*(Supplement C), 402–410. https://doi.org/10.1016/j.electacta.2014.12.126.

Ratte, H. T. (1999). Bioaccumulation and toxicity of silver compounds: A review. *Environmental Toxicology and Chemistry*, *18*(1), 89–108. https://doi.org/10.1002/etc.5620180112.

Rice, N. C., Rauscher, N. A., Langston, J. L. & Myers, T. M. (2018). Behavioral toxicity of sodium cyanide following oral ingestion in rats: Dose-dependent onset, severity, survival, and recovery. *Food and Chemical Toxicology*, *114*, 145–154.

Sadovnikov, S. I., Gusev, A. I. & Rempel, A. A. (2015). Artificial silver sulfide Ag2S: Crystal structure and particle size in deposited powders. *Superlattices and Microstructures*, *83*, 35–47. https://doi.org/10.1016/J.SPMI.2015.03.024.

Sasikumar, T. & Ilanchelian, M. (2020). Colorimetric and visual detection of cyanide ions based on the morphological transformation of gold nanobipyramids into gold nanoparticles. *New Journal of Chemistry*, *44*(12), 4713–4718. https://doi.org/10.1039/c9nj05929f.

Selvakannan, P. R., Swami, A., Srisathiyanarayanan, D., Shirude, P. S., Pasricha, R., Mandale, A. B. & Sastry, M. (2004). Synthesis of aqueous Au core-Ag shell nanoparticles using tyrosine as a pH-dependent reducing agent and assembling phase-transferred silver nanoparticles at the air-water interface. *Langmuir*, *20*(18), 7825–7836. https://doi.org/10.1021/la049258j.

Shamsipur, M., Karimi, Z. & Amouzadeh Tabrizi, M. (2017). A novel electrochemical cyanide sensor using gold nanoparticles decorated carbon ceramic electrode. *Microchemical Journal*, *133*, 485–489. https://doi.org/10.1016/j.microc.2017.04.017.

Singh, H. B., Wasi, N. & Mehra, M. C. (2006). *International Journal of Environmental Analytical Chemistry Detection and Determination of Cyanide — A Review*. Detection and Determination of Cyanida-A Review, *26*(February 2015), 37–41.

Stradiotto, N. R., Yamanaka, H. & Zanoni, M. V. B. (2003). Electrochemical sensors: a powerful tool in analytical chemistry. *Journal of the Brazilian Chemical Society*, *14*(2), 159–173. https://doi.org/10.1590/S0103-50532003000200003.

Supan, K. E., Robert, C., Miller, M. J., Warrender, J. M. & Bartolucci, S. F. (2017). Thermal degradation of MWCNT/polypropylene nanocomposites: A comparison of TGA and laser pulse heating. *Polymer Degradation and Stability*, *141*, 41–44. https://doi.org/10.1016/j.polymdegradstab.2017.05.006.

Surleva, A., Zaharia, M., Ion, L., Gradinaru, R. V., Drochioiu, G. & Mangalagiu, I. (2013). Ninhydrin-based spectrophotometric assays of trace cyanide. *Acta Chemica Iasi*, *21*(1), 57–70. https://doi.org/10.2478/achi-2013-0006.

Suzuki, T., Hioki, A. & Kurahashi, M. (2003). Development of a method for estimating an accurate equivalence point in nickel titration of cyanide ions. *Analytica Chimica Acta*, *476*(1), 159–165. https://doi.org/10.1016/S0003-2670(02)01362-4.

Taheri, A., Noroozifar, M. & Khorasani-Motlagh, M. (2009). Investigation of a new electrochemical cyanide sensor based on Ag nanoparticles embedded in a three-dimensional sol–gel. *Journal of Electroanalytical Chemistry*, *628*(1–2), 48–54. https://doi.org/10.1016/j.jelechem.2009.01.003.

Toh, S., Batchelor-Mcauley, C., Tschulik, K., Uhlemann, M., Crossley, A. & Compton, R. G. (n.d.). *The anodic stripping voltammetry of nanoparticles: electrochemical evidence for the surface agglomeration of silver nanoparticles*. https://doi.org/10.1039/c3nr00898c.

Wang, S., Lei, Y., Zhang, Y., Tang, J., Shen, G. & Yu, R. (2010). Hydroxyapatite nanoarray-based cyanide biosensor. *Analytical Biochemistry*, *398*(2), 191–197. https://doi.org/10.1016/J.AB.2009.11.029.

Wang, W., Gu, B., Liang, L. & Hamilton, W. (2003). Fabrication of Two- and Three-Dimensional Silica Nanocolloidal Particle Arrays. *The Journal of Physical Chemistry B*, *107*(15), 3400–3404. https://doi.org/10.1021/jp0221800.

Wilson, E. & Islam, M. F. (2015). Ultracompressible, High-Rate Supercapacitors from Graphene-Coated Carbon Nanotube Aerogels. *ACS Applied Materials & Interfaces*, *7*(9), 5612–5618. https://doi.org/10.1021/acsami.5b01384.

Wu, B., Kuang, Y., Zhang, X. & Chen, J. (2011). Noble metal nanoparticles/carbon nanotubes nanohybrids: Synthesis and applications. *Nano Today*, *6*(1), 75–90. https://doi.org/10.1016/J.NANTOD.2010.12.008.

Xing, W., Huang, C. C., Zhuo, S. P., Yuan, X., Wang, G. Q., Hulicova-Jurcakova, D., Yan, Z. F. & Lu, G. Q. (2009). Hierarchical porous carbons with high performance for supercapacitor electrodes. *Carbon*, *47*(7), 1715–1722. https://doi.org/10.1016/J.CARBON.2009.02.024.

Xu, F., Cai, R., Zeng, Q., Zou, C., Wu, D., Li, F., Lu, X., Liang, Y. & Fu, R. (2011). Fast ion transport and high capacitance of polystyrene-based hierarchical porous carbon electrode material for supercapacitors. *J. Mater. Chem.*, *21*(6), 1970–1976. https://doi.org/10.1039/C0JM02044C.

Xu, Z., Chen, X., Kim, H. N. & Yoon, J. (2010). Sensors for the optical detection of cyanide ion. *Chemical Society Reviews*, *39*(1), 127–137. https://doi.org/10.1039/b907368j.

Yang, T., Xu, J., Lu, L., Zhu, X., Gao, Y., Xing, H., Yu, Y., Ding, W. & Liu, Z. (2016). Copper nanoparticle/graphene oxide/single wall carbon nanotube hybrid materials as electrochemical sensing platform for nonenzymatic glucose detection. *Journal of Electroanalytical Chemistry*, *761*(Supplement C), 118–124. https://doi.org/10.1016/j.jelechem.2015.12.015.

You, B., Jiang, J. & Fan, S. (2014). Three-Dimensional Hierarchically Porous All-Carbon Foams for Supercapacitor. *ACS Applied Materials & Interfaces*, *6*(17), 15302–15308. https://doi.org/10.1021/am503783t.

Yu, L., Shearer, C. & Shapter, J. (2016). Recent Development of Carbon Nanotube Transparent Conductive Films. *Chemical Reviews*, *116*(22), 13413–13453. https://doi.org/10.1021/acs.chemrev.6b00179.

Yu, Y., Shu, T., Yu, B., Deng, Y., Fu, C., Gao, Y., Dong, C. & Ruan, Y. (2018). A novel turn-on fluorescent probe for cyanide detection in aqueous media based on a BODIPY-hemicyanine conjugate. *Sensors and Actuators, B: Chemical*, *255*, 3170–3178. https://doi.org/10.1016/j.snb.2017.09.142.

Zaheer, Z. & Rafiuddin. (2012). Silver nanoparticles formation using tyrosine in presence cetyltrimethylammonium bromide. *Colloids and Surfaces B: Biointerfaces*, *89*(1), 211–215. https://doi.org/10.1016/J.COLSURFB.2011.09.013.

Zelder, F. H. & Männel-Croisé, C. (2009). Recent advances in the colorimetric detection of cyanide. *Chimia*, *63*(1–2), 58–62. https://doi.org/10.2533/chimia.2009.58.

Zhang, Han, Xin, X., Sun, J., Zhao, L., Shen, J., Song, Z. & Yuan, S. (2016). Self-assembled chiral helical nanofibers by amphiphilic dipeptide derived from d- or l-threonine and application as a template for the synthesis of Au and Ag nanoparticles. *Journal of Colloid and Interface Science*, *484*, 97–106. https://doi.org/10.1016/J.JCIS.2016.08.052.

Zhang, Hongyuan, Sun, D. & Cao, T. (2020a). Electrochemical Sensor Based on Silver Nanoparticles/Multi-walled Carbon Nanotubes Modified Glassy Carbon Electrode to Detect Cyanide in Food Products. *Int. J. Electrochem. Sci*, *15*, 3434–3444. https://doi.org/10.20964/2020.04.32.

Zhang, Hongyuan, Sun, D. & Cao, T. (2020b). Electrochemical Sensor Based on Silver Nanoparticles/Multi-walled Carbon Nanotubes Modified Glassy Carbon Electrode to Detect Cyanide in Food Products. *Int. J. Electrochem. Sci*, *15*, 3434–3444. https://doi.org/10.20964/2020.04.32.

Zhang, Q., Maddukuri, N. & Gong, M. (2015). A direct and rapid method to determine cyanide in urine by capillary electrophoresis. *Journal of Chromatography A*, *1414*, 158–162. https://doi.org/10.1016/J.CHROMA.2015.08.050.

Chapter 2

A New Set of Thermochemically Stable Nitrile and Isonitrile Insertion Compounds with their Possible Trapping at Ambient Temperature

Gourhari Jana[1], Ranita Pal[2] and Pratim Kumar Chattaraj[1,3,*]

[1]Department of Chemistry, Indian Institute of Technology Bombay, Mumbai, India
[2]Advanced Technology Development Centre,
Indian Institute of Technology Kharagpur, India.
[3]Department of Chemistry, Indian Institute of Technology Kharagpur,
West Bengal, India

Abstract

Compounds containing cyanide (–CN) and isocyanide (–NC) functional groups are of utmost importance in chemistry owing to their prominent astronomical applications including their extensive involvement in star-formation, cold clouds formation, and in circumstellar shells. More than 200 molecules having –CN functional group are detected in interstellar medium and around 40 of them are synthesized in the laboratory as well. A number of alkali, alkaline earth, and transition metal containing cyanides and isocyanides such as NaCN, KCN, MgCN, MgNC, CaNC, AlNC, SiCN, SiNC, FeCN, FeNC and a list of other interstellar species including more complex molecules like isopropylcyanide are detected astronomically.

* Corresponding Author's E-mail: pkc@chem.iitkgp.ac.in.

In: Cyanide: Occurrence, Applications and Toxicity
Editor: Bill M. Torres
ISBN: 978-1-68507-619-1

In this mini-review, we briefly epitomize our works to date on theoretically predicted cyanide compounds having strong bonding units which can be synthesized under roughly ambient conditions. Our studied complexes include MNgCN/MNgNC, LMCN/LMNC, and NCNgNSi where Ng = Xe, Rn; M = Cu, Ag, Au; and L = C_2H_2, C_2H_4, CO, N_2, NH_3, H_2O, H_2S, and 1,3-dimethylimidazole (DMI). They have thermochemical stability concerning most of the dissociation paths, and their kinetic stability with respect to those channels which are thermodynamically spontaneous indicate their possible experimental realization. In combination with the structure and stability, our quantum chemical study also discusses the bonding senario in these compounds with the help of standard theoretical techniques like natural bond orbital, energy decomposition and electron density analyses. We also carefully examine and discuss possible ways of trapping the transient metal isocyanide isomer by introducing suitable ligands in an attempt to increase their kinetic stability. This report on cyanide and isocyanide compounds from the perspective of density functional theory along with high-level computations using coupled-cluster theory will guide experimentalists with information on structures, energetics, stability and, nature of chemical bonding which are essentially unknown.

1. Introduction

The astronomical importance of cyanide and isocyanide compounds are well-known and extensively reported in the literature [1-8]. The interstellar space is rich in cyanide compounds, especially metal cyanides like NaCN [9], MgCN [10], KCN [11], SiCN [12], and FeCN [10]. Heavy metal cyanides have important contribution towards controlling the cloud evolution [13]. Extensive study on the isomerization between cyanides and isocyanides has revealed that in certain cases, like those of the uranium compounds, the isocyanide isomer is preferred [14-19]. Whereas in the cases of transition metals, Sc-Fe (except Cr) seem to prefer the MNC isomer, and Co-Zn including Cr favours the MCN isomer [20]. The cyanide and isocyanide isomers of V, Mn, and Fe are very close in energy, and the energy difference between ionic cyanide and isocyanide, (e.g., LiCN and LiNC) is quite negligible. In comparison to organic isonitriles (RNC), nitriles (RCN) are thermodynamically more stable. It is noted that generally the main-group cyanide/isocyanide compounds prefer to exist in the form of cyanide due to their more covalent nature.

The cyanides of alkaline-earth-metal are very rarely explored. The abundance of trimethylsilyl isocyanide (Me_3SiNC) is reported to be shown in significant quantity than its cyanide analogue (Me_3SiCN). Interestingly, both experimental observation and theoretical findings suggest that the halide cyanide/isocyanide ratio (X-CN: X-NC) rises with the rise in the electronegativity of X and an explanation can be made that the XCN is becoming increasingly favourable for covalently bound CN groups. Metal cyanides, MCN (where M = coinage metals: Cu, Ag, Au) were reported to be experimentally found in the gas phase, [21, 22] but their isocyanide isomers were not. The extremely low kinetic energy barrier of the MNC → MCN isomerization process causes the transient MNC to easily convert to the more stable MCN isomer. Insertion of noble gas (Ng) within the M-C bond has proven to increase the activation barrier, as reported by Jana et al. [23].

Previously, noble gases were believed to exhibit little to no reactivity, and hence were also known by the term 'inert gases'. Extensive research on this field has gradually contributed towards dispelling this notion. Discovery of noble gases started in the late 1800s but it wasn't until 1962 that the first noble gas compound, $Xe^+[PtF_6]^-$, was discovered by Neil Bartlett [24]. That's when the research in this field gained real momentum since this discovery proved that these gases can form bonds, provided they have compatible partners and appropriate conditions. Most commonly researched Ng compounds contain the Ng-C bonds [25-34], followed by those containing Ng-N bonds [35-44]. Compounds containing both Ng-C and Ng-N bonds, however, was unheard of until recently. Pan et al. [45], in 2018, inserted Ng atoms within the NC-NSi bond of an experimentally identified compound, silaisocyanogen (NCNSi) [26], to form NCNgNSi (Ng = Xe, Rn) and studied their thermochemical and kinetic stability. Coupled-cluster calculations along all possible dissociation channels of NCNgNSi revealed that they are thermochemically stable except along the Ng-release channel where Ng and CNSiN moieties are produced. This spontaneous channel is, however, kinetically hindered by an activation free energy barrier ($\Delta G^{\neq}$) of 37.0 and 39.3 kcal/mol, for Ng = Xe and Rn, respectively, suggesting their possible stability at ambient temperature. An interesting aspect is noted while studying the Ng inserted NCNSi compounds. While this parent compound, i.e., NCNSi, was spectroscopically detected, its other isomer (i.e., CNSiN) was not. However, our computations show that at elevated temperatures the release of Ng atoms from NCNgNSi can lead to the formation of CNSiN.

This essentially suggests that the Ng insertion within the NC-NSi bond of NCNSi has somehow provided enough space for the CN and NSi moieties to simultaneously flip at higher temperatures to form the higher energy isomer, CNSiN. A detailed analysis of the isomerization process in presence and absence of Ng is thus delineated here along with a discussion on the bonding properties of these compounds.

Other 'noble' elements in the periodic table, namely the noble metals, also constitute an area of high research interest. The possibility of bonding between these two noble entities, i.e., noble gases and noble metals, is highly fascinating since both of these moieties are considered to exhibit low chemical reactivity. In the year 1995, $NgAu^+$ and $NgAuNg^+$ species [46] were predicted to be viable by Pyykkö. Shortly thereafter, in 1998, the Xe analogues of the aforementioned compounds were experimentally detected by Schröder et al. [47] using mass spectroscopy. Several molecular species with bonds between noble gases and noble metals are reported in the literature. In the years 2005 and 2006, the metastable MNgX (M = Cu, Ag, Au; Ng = Kr, Xe; and X = F, OH) [48, 49] compounds were reported by Ghanty. In one of our previous works, we have reported for the first time, MNgCCH (M = Cu, Ag, Au; Ng = Xe, Rn), [50] which contains the M-Ng-C bonding unit. As previously mentioned, research on compounds containing Ng-C bonds are abundantly available in the literature. Among them, it is quite evident that the use of –CN moiety provides a higher stability to the Ng inserted compounds. Here we discuss the stability and bonding of an Ng inserted MCN species (MNgCN; M = Cu, Ag, Au) *via* coupled-cluster computations, studied by Jana et al. [23]. The Ng release paths, MNgCN $\rightarrow$ Ng + MCN/MNC, although exergonic, are kinetically restricted, and are associated with MNgCN $\rightarrow$ NgMCN/NgMNC isomerization processes. The isomerization process producing NgMCN occurs *via* an intermediate MNgNC by a 180° rotation of the –CN moeity. The one producing NgMNC, is a single-step isomerization although NgMNC further gets converted into NgMCN isomer. Although these isomeric transformations leading to the most stable structure are more complex than other Ng-inserted compounds, they enable the efficient trapping of the transient MNC as NgMNC isomer which has a higher kinetic stability. The corresponding kinetic barrier of MNC is too low for the species to be detected experimentally. This Ng insertion process has provided a way to trap the isocyanide isomer by increasing the activation free energy barrier.

Another method of increasing the said kinetic barrier in an attempt to trap the elusive MNC isomer was attempted by our group [51] where we have used some interacting small gas molecules as ligands and investigated the LMNC → LMCN isomerization path (L = C_2H_2, C_2H_4, CO, N_2, NH_3, H_2O, H_2S, and 1,3-dimethylimidazole (DMI)), which is also discussed in this mini-review article.

2. Computational Details

The geometry optimizations and frequency analysis of MNgCN, NCNgNSi and their possible isomers are performed at the CCSD(T) [52], MP2, [53] and MPW1B95, [54] levels of theory. Basis sets, cc-pVTZ-PP and def2-TZVPP are selected for the aforementioned studies [55-57]. The ωB97X-D functional [58] and MP2 [53] level are used for the study of LMNC/LMCN isomerization with basis set, cc-pVTZ-PP. The basis set cc-pVTZ [55] is considered for H, C, N, O, Si, S and Ar atoms. Relativistic effective core potentials (RECPs), ECP10MDF for Cu and Kr, ECP28MDF for Ag and Xe, and ECP60MDF for Au and Rn, are considered [59]. The zero point corrected energy (ZPE), enthalpy and Gibbs free energies are obtained from the frequency calculation. For predicting the TSs of the isomerization LMNC → LMCN, we have used ωB97X-D/cc-pVTZ/cc-pVTZ-PP level since MP2 overestimates free energy barrier and bond energy [60-62]. Whereas, MPW1B95/cc-pVTZ-PP level and MPW1B95/def2-TZVPP level are employed for MNgCN and NCNgNSi compounds to avoid high computational cost of CCSD(T). MPW1B95 functional is reported as benchmark standard which reproduces experimental and/or CCSD(T) values to a very good approximation.

NBO [63] analysis, to evaluate the natural charge, is performed and Wiberg bond indices (WBI) [64] are calculated using the Gaussian package [65]. The electron density analysis [66] with all-electron WTBS for the coinage metals and the Ng atoms [67] is performed in Multiwfn [68]. The *n*-center-two-electron (*n*c-2e) bonds are detected and analyzed using the AdNDP method.

Energy decomposition analysis (EDA) at PBE-D3(BJ) [69-71] /TZ2P (for MNgCN and LMCN) and PBE-D3(BJ) [69-72] /TZ2P (for NCNgNSi) levels is performed in ADF 2013.01 package [73].

The zeroth-order regular approximation (ZORA) [74-77] takes care of the scalar relativistic effects in the heavy elements. In EDA, the sum of electrostatic, orbital, dispersion, and Pauli repulsion make up the total interaction energy (ΔE_{int}) between two fragments, and can be represented as,

$$\Delta E_{int} = \Delta E_{Pauli} + \Delta V_{elstat} + \Delta E_{orb} + \Delta E_{disp} \quad (1)$$

Local reactivity descriptors at the atomic centers are calculated within the domain of Conceptual DFT (CDFT). Fukui function ($f(r)$) [78], and its dual descriptor $\Delta f(r)$ [79, 80] are calculated as

$$f_k^{\,+} = q_k(N+1) - q_k(N) \text{ nucleophilic attack} \quad (2)$$

$$f_k^{\,-} = q_k(N) - q_k(N-1) \text{ electrophilic attack} \quad (3)$$

$$\Delta f_k(r) \cong [(f_k^{\,+}(r)) - (f_k^{\,-}(r))] \quad (4)$$

q_k being the natural population at the kth center of a molecule with N electrons

Eectrophilicity (ω) [81] based multiphilic descriptor, $\Delta\omega(r)$ [82], are evaluated as

$$\omega = \frac{\mu^2}{2\eta} \quad (5)$$

$$\Delta\omega_k = [\omega_k^{+} - \omega_k^{-}] \quad (6)$$

where,

$$\omega_k^{\alpha} = \omega f_k^{\alpha} \ (\alpha = +, -) \quad (7)$$

μ and η being the chemical potential and chemical hardness, respectively, which in turn are defined as follows [83-87]:

$$\mu = \left(\frac{\partial E}{\partial N}\right)_{v(r)} \approx -\frac{IP+EA}{2} \approx \frac{\varepsilon_{HOMO}+\varepsilon_{LUMO}}{2} \quad (8)$$

$$\eta = \left(\frac{\partial \mu}{\partial N}\right)_{v(r)} = \left(\frac{\partial^2 E}{\partial N^2}\right)_{v(r)} \approx IP - EA \approx \varepsilon_{LUMO} - \varepsilon_{HOMO} \quad (9)$$

where EA, IP, ε_{HOMO} and ε_{LUMO} are electron affinity, ionization potential, highest occupied and lowest unoccupied molecular orbital energies, respectively.

An attempt to validate the electronic structure principles are made along the LMNC → LMCN isomerization path. Minimum electrophilicity (MEP) [88-91] and maximum hardness (MHP) [92-95] principles state that along a reaction path or in a system, the minimum electrophilicity and maximum hardness should correspond to the equilibrium geometry, whereas the TS geometry should have maximum electrophilicity and minimum hardness values. More details regarding the analysis on the principles are provided later in the chapter.

3. Results and Discussion

Among the previously reported compounds, it is evident that compounds containing cyanide fragment adjacent to the Ng atom are quite effective in stabilizing Ng inserted molecules. Here we have discussed certain systems containing -CN functionality having great practical interests in environmental, industrial, and medical applications, astronomical importance, and could be experimentally synthesized. The compounds with molecular formulas, MNgCN (M = Cu, Ag, Au), LMCN/NC (L = CO, H_2O, H_2S, N_2, NH_3, DMI, C_2H_2, and C_2H_4), NCNgNSi, and their isomers are investigated here along with their electronic properties. It is noteworthy to mention that monomeric MCN (M = Cu, Ag, Au) and NCNSi molecules are considered as the parent moieties to make our attempt more realistic, since they are detected in the gas phase.

3.1. MNgCN (Ng = Xe, Rn) and LMCN

3.1.1. Structure and Stability

The experimentally detected MCN is more stable by 13.3, 14.8, and 24.4 kcal/mol than MNC, for Cu, Ag, and Au analogues, respectively. This instability of MNC relative to the MCN is comparable to those in the experimentally reported CH_3NC (24.4 kcal/mol) and HNC (14.7 kcal/mol).

Since the stability of an isomer is associated with the free energy barrier ($\Delta G^{\ddagger}$) of its isomeric transformation, a reasonably high $\Delta G^{\ddagger}$ would make it difficult for the higher-energy isomer to convert to its lowest energy isomer, hence making it viable at ambient temperatures. While the $\Delta G^{\ddagger}$ values are high enough for the isomeric transformation of CH_3NC and HNC, in the case of AuNC, it has a moderate value ($\Delta G^{\ddagger}$ = 7.1 kcal/mol, could be viable at low temperatures), and for Cu and Ag analogues, they are very low (1.4 and 1.9 kcal/mol, respectively). It suggests that the latter two analogues can easily convert to their respective cyanide isomers, CuCN and AgCN. In other words, their isocyanide isomers, MNC, have neither thermodynamic nor kinetic stablility. Yet, their detection at low temperatures cannot be ruled out. This explains why the MNC form is still elusive and the cyanide form, MCN, was found exclusively experimentally. To trap the MNC and observe its viability at room temperature, we have performed a couple of studies, the first one dealing with the insertion of Ng in the M-C bond, while the second one uses small interacting ligands (L= CO, H_2O, H_2S, N_2, NH_3, DMI, C_2H_2, and C_2H_4) to bind with the M center. We have introduced Ng into the M-C bond (the obvious choice between M-C and C≡N) of the experimentally detected MCN (M = Cu, Ag, Au).

The structural analysis, stability, and bonding patterns of MNgCN (Ng = Xe, Rn) complexes are explored *via* a high level of *ab-initio* theory i.e, coupled-cluster method. Minimum energy structures of the insertion isomer, MNgCN (Ng = Xe, Rn), adopt linear geometries having $^1\Sigma_g$ electronic state. The non-insertion isomers, NgMCN and NgMNC, are the lower energy isomers (Figure 1) indicated by the coupled-cluster theory of computations. The insertion isocyanide isomer, MNgNC, has higher energy than MNgCN. The thermochemical stability is revealed by examining possible ways the compounds can undergo dissociation. From this analysis it is found that all two- and three-body (2B and 3B) dissociations are endergonic except MNgCN $\rightarrow$ Ng + MNC and MNgCN $\rightarrow$ Ng + MCN channels (Table 1). These two 2B dissociations occur *via* the internal isomerization, MNgCN $\rightarrow$ NgMNC and MNgCN $\rightarrow$ NgMCN, respectively. Although the dissociation channels are exergonic, these internal rotations of the insertion to non-insertion isomers are kinetically protected by a considerable amount of activation free energy barriers, indicating the metastability of these compounds.

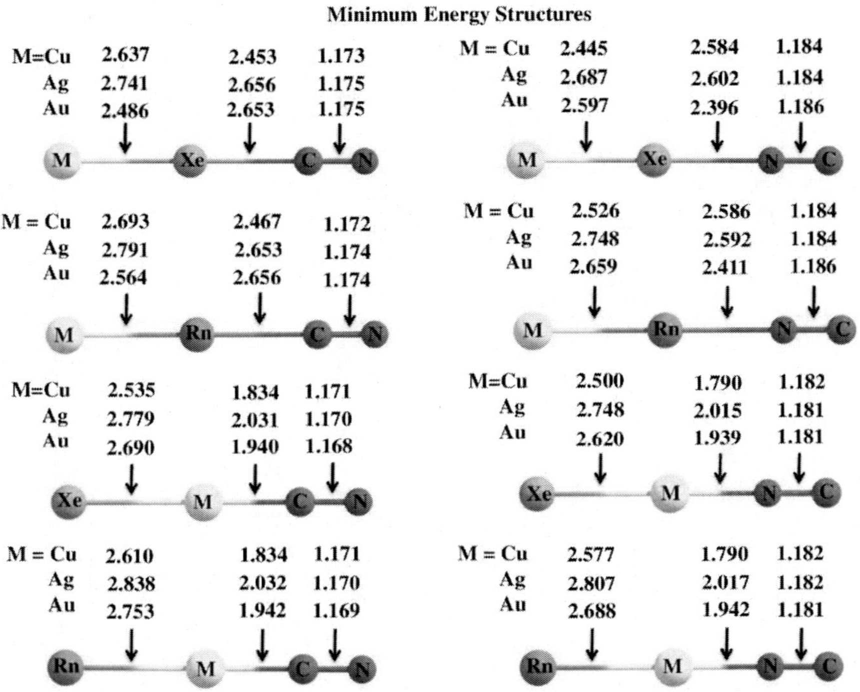

Figure 1. Geometries of MNgCN, MNgNC, NgMCN and NgMNC optimized at the CCSD(T)/cc-pVTZ/cc-pVTZ-PP level (bond distances (*r*) in Å). (Adapted from Jana et al., [23] with permission from PCCP Owner Societies.).

The binding of Ng atoms (Ng = Xe, Rn) at the terminal metal center of MNC, moderately increases the activation free energy barrier (NgMNC → Ng + MNC has energy barrier values ranging within 9.4-19.3 kcal/mol) causing easy trapping of the MNC with Ng. Hence, it is proposed that NgMNC may be viable at room temperature. Interestingly, MNgCN → NgMCN occurs by the 180° flipping of the –CN moeity *w.r.t.* the M-Ng moiety to form the intermediate compound, MNgNC, which then transforms to the non-insertion isomer, NgMCN. Whereas, the MNgCN → NgMNC conversion is a single-step process, although, NgMNC further crosses a low barrier and transforms to NgMCN (two-step process for the Cu analogue, and one-step in the cases of Ag and Au).

Table 1. ZPE corrected dissociation energy (D_0), and change in free energy at 298 K (ΔG) in kcal/mol for possible dissociation paths of MNgCN species at the CCSD(T)/cc-pVTZ/cc-pVTZ-PP level. (Reprinted from Jana et al., [23] with permission from PCCP Owner Societies.)

Processes		D_0		ΔG	
		Xe	**Rn**	**Xe**	**Rn**
MNgCN → M + Ng + CN	Cu	47.2	55.8	33.7	42.1
	Ag	15.2	24.1	1.5	10.3
	Au	53.7	64.8	39.4	50.3
MNgCN → M^+ + Ng + CN^-	Cu	112.2	120.9	99.5	107.9
	Ag	102.7	111.6	89.8	98.6
	Au	146.1	157.1	132.6	143.4
MNgCN → M^- + Ng + CN^+	Cu	315.3	323.9	302.6	311.0
	Ag	306.9	315.8	294.0	302.8
	Au	285.3	296.4	271.8	282.7
MNgCN → MNg^+ + CN^-	Cu	92.6	98.4	83.4	88.9
	Ag	89.3	95.4	79.8	85.7
	Au	119.0	124.8	108.9	114.5
MNgCN → M^+ + $NgCN^-$	Cu	110.5	118.4	101.1	108.9
	Ag	100.9	109.1	91.4	99.6
	Au	144.5	154.8	134.2	144.5
MNgCN → M^- + $NgCN^+$	Cu	216.4	214.5	201.9	200.2
	Ag	208.0	206.4	201.9	200.2
	Au	186.4	187.0	179.7	180.1
MNgCN → Ng + MCN	Cu	-78.8	-70.1	-84.4	-76.0
	Ag	-69.6	-60.7	-75.7	-66.9
	Au	-72.4	-61.3	-78.6	-67.8
MNgCN → Ng + MNC	Cu	-66.8	-58.2	-73.0	-64.6
	Ag	-56.1	-47.2	-62.8	-54.0
	Au	-46.5	-35.4	-53.4	-42.5
MNgCN → NgMCN	Cu	-91.0	-83.5	-90.2	-82.8
	Ag	-78.2	-70.6	-77.9	-70.5
	Au	-87.0	-77.8	-86.4	-77.4
NgMCN → Ng + MCN	Cu	12.5	13.6	5.8	6.8
	Ag	8.7	10.1	2.2	3.5
	Au	14.9	16.7	7.7	9.6
MNgCN → NgMNC	Cu	-81.1	-73.6	-80.6	-73.1
	Ag	-65.5	-58.0	-65.5	-58.2
	Au	-65.8	-56.7	-65.5	-56.5
NgMNC → Ng + MNC	Cu	14.4	15.4	7.6	8.5
	Ag	9.4	10.8	2.7	4.1
	Au	19.3	21.2	12.0	13.9
MNgCN → MNgNC	Cu	4.6	5.2	4.6	3.7
	Ag	5.5	5.9	3.8	4.1
	Au	10.6	10.3	9.8	9.8

The overall isomerization process for the Au analogue is presented in Figure 2. This isomeric conversion into the most stable isomer is more complex than in the case of other Ng-inserted compounds. Interestingly, this is the reason of the increased activation free energy barrier which in turn increases the kinetic stability of the isocyanide isomers. Again, the Ng-release processes of the non-insertion compounds, NgMCN and NgMNC are endergonic following Au > Cu > Ag order for a given Ng. It is thus evident that MNC has a higher Ng binding ability than MCN.

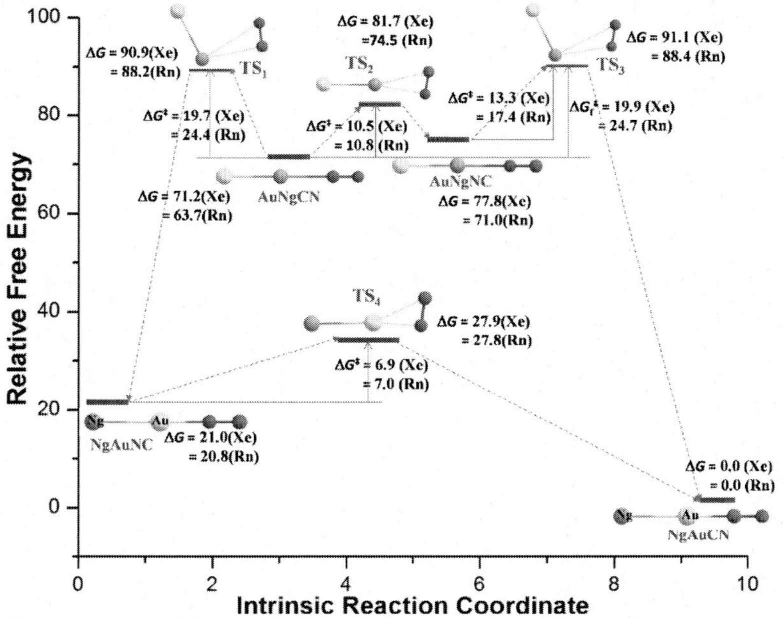

Figure 2. The overall isomeric transformation in AuNgCN species reported at the MPW1B95/cc-pVTZ/cc-pVTZ-PP level (Energies are provided in kcal/mol unit). (Adapted from Jana et al., [23] with permission from PCCP Owner Societies.).

Another attempt to improve the kinetic stability of the MNC isomer involves the investigation of the isomerization process, MNC → MCN, in the presence of different interacting ligands like small gas molecules (CO, H_2O, H_2S, N_2, NH_3, DMI, C_2H_2, and C_2H_4). The LMCN/NC subjected to MP2 and DFT based calculations. Figure 3 depicts the geometries of all the cyanide compounds and their respective TS for the isomerization process optimized at ωB97X-D/cc-pVTZ/cc-pVTZ-PP level.

To determine the viability of these complexes, different possible dissociation channels and their corresponding thermochemical parameters such as change in Gibbs free energy (ΔG) and bond dissociation energy (D_0) are investigated (see Table A-1). There are five two-body (2B) and three three-body (3B) paths of possible dissociation. among which, barring MNgCN → Ng + MNC and MNgCN → Ng + MCN channels, all paths are highly endergonic (ΔG = 79.8-311.0 kcal/mol). The dissociation, MNgCN → M + Ng + CN is endergonic with ΔG ranging from 1.5 to 39.4 kcal/mol, and 10.3 to 50.3 kcal/mol for the Xe and Rn analogues, respectively. The LMNC → LMCN isomerization is exergonic by 7.5-11.6 for Cu, 11.7-15.1 Ag, and 16.2-25.0 kcal/mol for Au (Figure 4 depicts those of the Au analogues at ωB97X-D level). The corresponding activation free energy barriers reveal the kinetic stability of MNC which is improved as 5.5-15.3 for Cu, 1.9-4.7 for Ag, 8.1-10.1 for Au (in kcal/mol) in the presence of the selected set of ligands.

3.1.2. Nature of Bonding in MNgCN

A detailed bonding situation is characterized *via* NBO, EDA, and AdNDP analyses. The M-Ng bonds are found to be covalent but have a significant electrostatic contribution, whereas the Ng-C bonds are shown to have predominantly electrostatic interaction.

NBO reveals that the Ng and M atoms in MNgCN acquire 0.22-0.55 |e| and 0.23-0.71 |e| natural charges, respectively. The net computed electronic charge on MNg is 0.78-0.93 |e| suggesting the molecular representation to be $(MNg)^+(CN)^-$. Au has less positive charge compared to Cu and Ag (see Table 2), whereas, the opposite trend is true for the Ng atoms in the respective complexes (i.e., they have higher positive charge in the Au analogues than in the Cu and Ag complexes). For a given Ng atom in the M-Ng bonds, the WBIs vary within 0.37-0.68 which follows the trend, Au > Cu > Ag. For a given M, the covalent character of the M-Xe is greater than M-Rn. The Ng-C bonds have appreciable electrostatic character judging by the low WBIs and the charge distribution. Comparing the M-Ng and Ng-C bond distances in MNgCN with the corresponding r_{cov} values, it is found that the difference is very small (0.024-0.151 Å) in the case of M-Ng, and large in case of Ng-C (0.297-0.596 Å). Also, the Au-C bond is shorter than Cu-C and Ag-C, further indicating higher covalency in the former case.

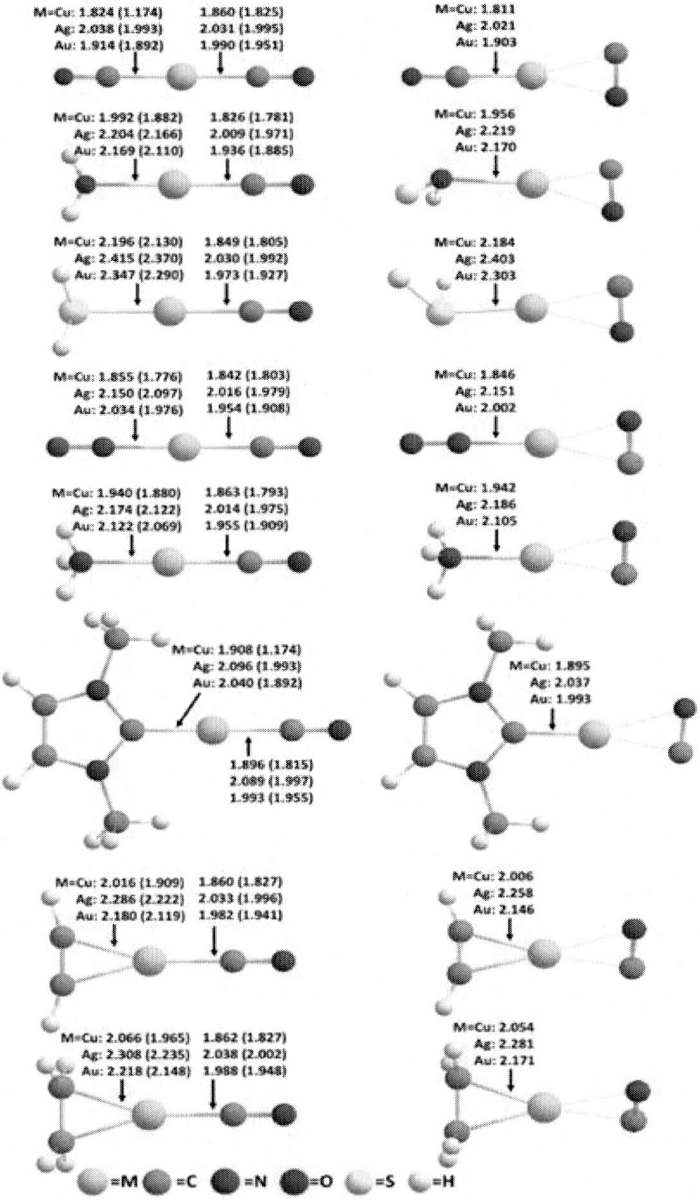

Figure 3. Geometries of LMCN complexes (M = Cu, Ag, Au; L= CO, H_2O, H_2S, N_2, NH_3, DMI, C_2H_2 and C_2H_4) along with their TS geometries at the ωB97X-D/cc-pVTZ/cc-pVTZ-PP level. The bond lengths (Å) within parentheses are calculated at MP2 level. (Reprinted from Pal et al. [51] with permission from Springer Nature. Copyright © 2019, Springer-Verlag GmbH Germany, part of Springer Nature).

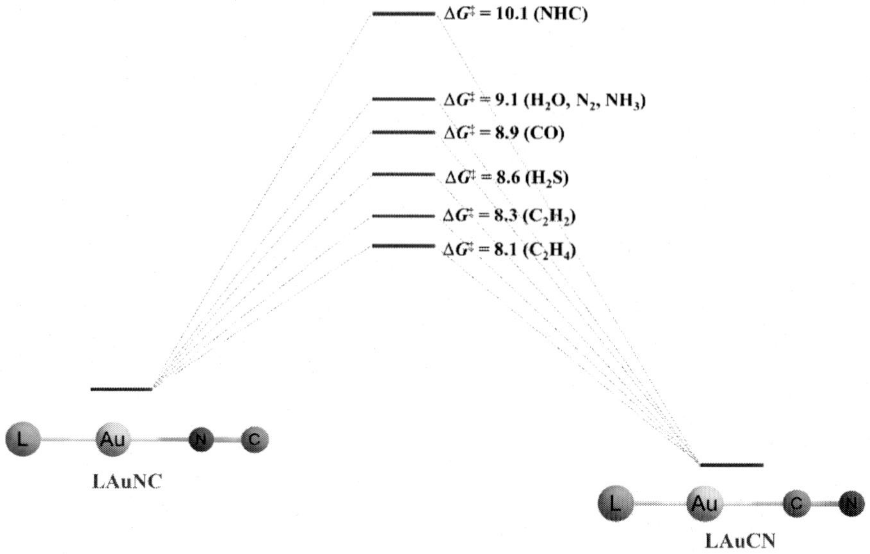

Figure 4. Activation free energy barrier of LAuCN (L= CO, H_2O, H_2S, N_2, NH_3, DMI, C_2H_2 and C_2H_4) complexes obtained at the ωB97X-D/cc-pVTZ/cc-pVTZ-PP level. (Energies are provided in kcal/mol unit). (Reprinted from Pal et al. [51] with permission from Springer Nature. Copyright © 2019, Springer-Verlag GmbH Germany, part of Springer Nature).

From the AdNDP results, we see that in MNgCN, the M center contains five LPs while the Ng center contains three LPs, one LP is located on the N atom (**Figure 5**). The -CN moeity retains its triple bond nature (σ-bond and two π-bonds) after Ng insertion. The localized 2c-2e σ-bond in M-Ng with occupation number (ON) =1.98-199 $|e|$, is polarized towards the Ng center. That of the Ng-C bond has ON = 1.82-1.91 $|e|$. Reasonably high ON dictates the identification of polarized 2c-2e Ng-C σ-bond. Therefore, an ionic interaction between MNg and CN fragments is expected judging by the charge distribution, although, an appreciable amount of covalent character also exists therein.

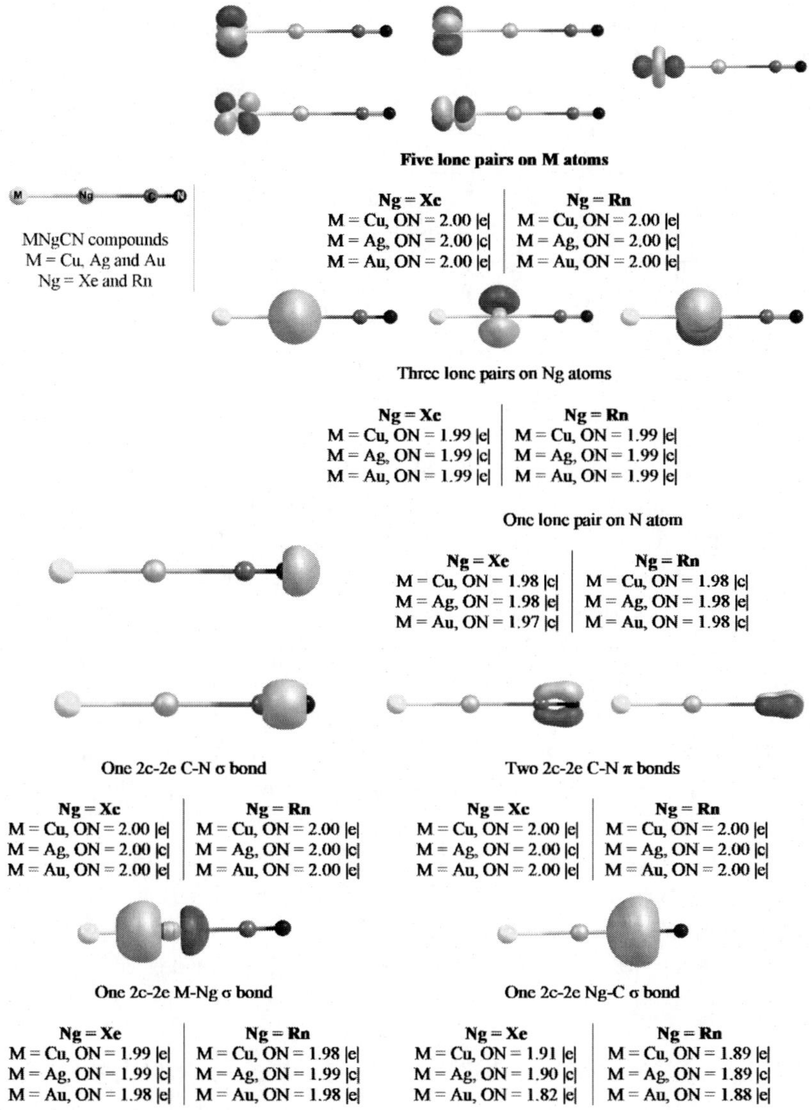

Figure 5. Bonding elements of MNgCN complexes (Ng = Xe, Rn) obtained from AdNDP analysis. (Reprinted from Jana et al., [23] with permission from PCCP Owner Societies.).

Table 2. The natural charges on Ng and M centers (q, au), WBI for Ng-M and Ng-C bonds of MNgCN molecules at the CCSD(T)/cc-pVTZ/cc-pVTZ-PP level. (Reprinted from Jana et al., [23] with permission from PCCP Owner Societies.)

Systems	q		WBI	
	Ng	M	Ng-M	Ng-C
CuXeCN	0.25	0.68	0.42	0.08
CuRnCN	0.32	0.60	0.50	0.09
AgXeCN	0.22	0.71	0.37	0.08
AgRnCN	0.29	0.61	0.46	0.10
AuXeCN	0.45	0.35	0.62	0.19
AuRnCN	0.55	0.23	0.68	0.21

The EDA results are presented to understand the bonding in further details. From the NPA charge analysis we inferred that both the neutral and ionic fragmentations are possible. Generally, when both the schemes are possible, the one with least ΔE_{orb} value turns out to be a better representative since a lower ΔE_{orb} value indicates that those fragments require a lower amount of alternation in its charge distribution in forming the electronic structure of the complex [96]. For the M-Ng bonds, the preferred scheme is the neutral one as it has significantly smaller ΔE_{orb} values than those of the ionic decomposition (Table 3). For the Ng-C bonds, however, the ionic has smaller ΔE_{orb} values, although the difference is quite low with those of the neutral scheme (Table 4). The M-Ng bonds cannot be considered as typical covalent bonds since they have a higher electrostatic contribution. Although the ionic scheme in the Ng-C bonds have ΔV_{elstat} as the highest contributing term, the orbital contribution is also considerable, which is even comparable to that of the neutral scheme. This is the main reason for describing the Ng-C bonds as covalent in the EDA analysis.

3.1.3. Nature of Bonding and CDFT Principles in LMCN Complexes

Similar studies performed on the LMCN/NC complexes reveal the partial covalent nature of the L–M bonds (especially for L = CO, DMI, and H_2S). For a particular ligand, Au has the highest binding ability, followed by Cu and Ag. A σ-donation from $HOMO_{(L)}$ to $LUMO_{(MCN)}$ along with both σ- and π-back donations from the MCN MOs to the vacant $d_{(M)}$ orbitals are the main stabilizing factors as revealed from the EDA-NOCV analysis.

The interactions within the L-M bonds are dominated by the ΔV_{elstat} term (55.4–77.4% toward the total attraction) which indicates a predominant electrostatic interaction therein. The transfer of charge and the pairwise orbital interactions between the interacting fragments are the main reasons behind the increased M-C/N bond length, which in turn facilitates the internal rotation.

The validity of the CDFT principles (MHP and MEP) [88-95] are investigated along the isomerization path, LAuNC → LAuCN, taking the largest (DMI) ligand as the case study. As per these principles, the LAuCN (most stable) isomer should correspond to higher η and lower ω compared to those in LAuNC. Again, both of these isomers should have higher η and lower ω than those of the TS. Although the trend in η is follows MHP, that of the ω must fulfil another criterion [94] to obey MEP. Both η and μ must attain a simultaneous maximum for ω to be a minimum, and vice versa. In the studied cases, η and μ shows opposite trend and nowhere along the isomerization path do they attain a simultaneous maximum or minimum (**Figure 6**). Thus, while the MHP is obeyed, the MEP is not followed in this isomerization process.

Table 3. The total interaction energy along with Pauli, electrostatic, orbital and dispersion energies within the M-Ng bond of MNgCN, obtained from EDA performed at the PBE-D3/QZ4P//CCSD(T)/cc-pVTZ/cc-pVTZ-PP level. (Reprinted from Jana et al., [23] with permission from PCCP Owner Societies.)

Systems	Fragments	ΔE_{int}	ΔE_{Pauli}	ΔV_{elstat}[a]	ΔE_{orb}[a]	ΔE_{disp}[a]
CuXeCN	Cu + XeCN	-34.1	87.3	-63.4(52.2)	-57.9(47.7)	-0.1(0.1)
CuRnCN	Cu + RnCN	-38.2	84.9	-66.1(53.7)	-56.9(46.2)	-0.1(0.1)
AgXeCN	Ag + XeCN	-27.6	63.2	-45.3(49.9)	-45.2(49.8)	-0.3(0.3)
AgRnCN	Ag + RnCN	-31.8	66.9	-51.6(52.3)	-46.8(47.4)	-0.3(0.3)
AuXeCN	Au + XeCN	-34.7	98.5	-71.6(53.8)	-61.4(46.0)	-0.3(0.2)
AuRnCN	Au + RnCN	-41.1	103.0	-80.5(55.9)	-63.3(43.9)	-0.3(0.2)
CuXeCN	Cu^+ + $XeCN^-$	-159.7	57.2	-138.1(63.7)	-78.7(36.3)	-0.1(0.1)
CuRnCN	Cu^+ + $RnCN^-$	-167.4	60.9	-147.9(64.8)	-80.2(35.2)	-0.1(0.1)
AgXeCN	Ag^+ + $XeCN^-$	-143.0	53.2	-128.2(65.3)	-67.8(34.5)	-0.3(0.1)
AgRnCN	Ag^+ + $RnCN^-$	-151.2	64.3	-145.0(67.3)	-70.1(32.6)	-0.3(0.1)
AuXeCN	Au^+ + $XeCN^-$	-200.2	113.0	-196.5(62.7)	-116.5(37.2)	-0.3(0.1)
AuRnCN	Au^+ + $RnCN^-$	-209.8	128.6	-219.6(64.9)	-118.4(35.0)	-0.3(0.1)

[a] The values in parentheses are percentage contribution toward the total attraction, $\Delta V_{elstat} + \Delta E_{orb} + \Delta E_{disp}$

Table 4. The total interaction energy along with Pauli, electrostatic, orbital and dispersion energies within the Ng-C bond of MNgCN, obtained from EDA performed at the PBE-D3/QZ4P//CCSD(T)/cc-pVTZ/cc-pVTZ-PP level. (Reprinted from Jana et al., [23] with permission from PCCP Owner Societies.)

Systems	Fragments	ΔE_{int}	ΔE_{Pauli}	$\Delta V_{elstat}{}^{a}$	$\Delta E_{orb}{}^{a}$	$\Delta E_{disp}{}^{a}$
CuXeCN	$CuXe^+ + CN^-$	-111.8	70.0	-126.5(69.6)	-54.8(30.2)	-0.4(0.2)
CuRnCN	$CuRn^+ + CN^-$	-116.0	80.0	-139.2(71.0)	-56.4(28.8)	-0.5(0.2)
AgXeCN	$AgXe^+ + CN^-$	-106.4	70.6	-121.7(68.7)	-54.9(31.0)	-0.4(0.2)
AgRnCN	$AgRn^+ + CN^-$	-110.7	82.2	-135.5(70.3)	-56.9(29.5)	-0.5(0.2)
AuXeCN	$AuXe^+ + CN^-$	-132.1	115.6	-160.4(64.8)	-86.8(35.0)	-0.5(0.2)
AuRnCN	$AuRn^+ + CN^-$	-136.1	127.5	-177.4(67.3)	-85.7(32.5)	-0.5(0.2)
CuXeCN	CuXe + CN	-76.3	34.1	-16.0(21.0)	-59.9(78.5)	-0.4(0.5)
CuRnCN	CuRn + CN	-89.5	41.0	-20.4(22.8)	-68.6(76.7)	-0.5(0.5)
AgXeCN	AgXe + CN	-73.3	34.7	-16.1(22.0)	-56.8(77.5)	-0.4(0.6)
AgRnCN	AgRn + CN	-87.9	42.5	-20.9(23.7)	-66.6(75.7)	-0.5(0.5)
AuXeCN	AuXe + CN	-101.8	61.9	-28.9(28.4)	-72.4(71.1)	-0.5(0.4)
AuRnCN	AuRn + CN	-118.2	70.3	-34.7(29.4)	-83.0(70.2)	-0.5(0.4)

[a] The values in parentheses are percentage contribution toward the total attraction, $\Delta V_{elstat} + \Delta E_{orb} + \Delta E_{disp}$.

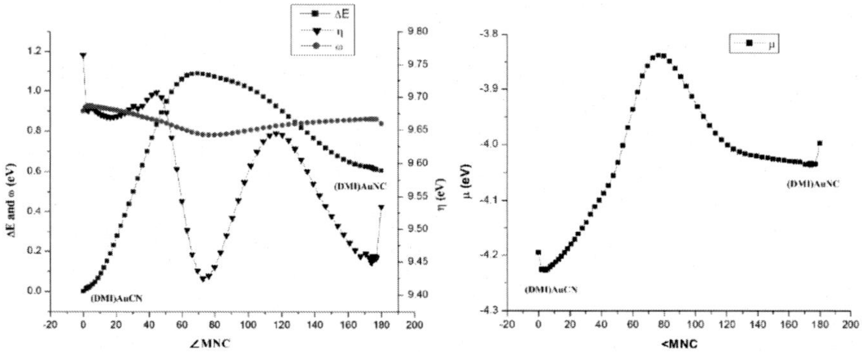

Figure 6. Variations in η, ω, μ and ΔE (in eV) with the variation of M-Ng-C bond angle for (DMI)AuNC → (DMI)AuCN isomerization path. (Reprinted from Pal et al. [51] with permission from Springer Nature. Copyright © 2019, Springer-Verlag GmbH Germany, part of Springer Nature).

To better understand the reactivity of the LMCN/NC complexes towards any electrophile/nucleophile, we have computed CDFT based local reactivity descriptors at each center of the complex taking N_2AuCN/N_2AuNC as the case study. Δf_k and $\Delta\omega_k$ calculated at each center are provided in Table 5. Positive value of both of these descriptors indicate that the center is prone to a nucleophilic attack, whereas both having negative values suggest that the center is prone to an electrophilic attack. While Au in the cyanide isomer prefers an electrophilic attack, that in the isocyanide isomer prefers a nucleophilic attack. Both of the N centers in the ligand prefer nucleophilic attack, In the cyanide isomer, the C and N centers of the cyanide moeity are prone to nucleophilic attack. In the TS, however, they change their reactivity (becomes prone towards electrophilic attack). Finally, in the isocyanide isomer, the N center of the –NC group favours electrophilic attack and the C center favours nucleophilic attack.

Table 5. CDFT based dual descriptor (Δf_k) and multiphilic descriptor ($\Delta\omega_k$) computed at the MP2/cc-PVTZ-PP level of theory. (Reprinted from Pal et al. [51] with permission from Springer Nature. Copyright © 2019, Springer-Verlag GmbH Germany, part of Springer Nature)

Compounds	**Dual descriptor/Multiphilic descriptor at Atomic Sites**				
	Δf_N	Δf_N	Δf_{Au}	$\Delta f_{C/N}$	$\Delta f_{N/C}$
N_2AuCN	0.163	0.366	-0.621	0.031	0.058
TS	0.209	0.313	0.045	-0.539	-0.028
N_2AuNC	0.224	0.342	0.139	-0.755	0.052
	$\Delta\omega_N$	$\Delta\omega_N$	$\Delta\omega_{Au}$	$\Delta\omega_{C/N}$	$\Delta\omega_{N/C}$
N_2AuCN	0.010	0.023	-0.038	0.002	0.004
TS	0.010	0.015	0.002	-0.027	-0.001
N_2AuNC	0.011	0.017	0.007	-0.038	0.003

3.2. NCNgNSi (Ng = Kr, Xe and Rn)

Astronomically important Silaisocyanogen, NCNSi, was quite recently detected *via* experimental identification. The possible isomeric transformation of NCNSi compound can occur to three different isomers such as NCSiN, CNSiN and CNNSi. Our theoretical investigation at MPW1B95/def2-TZVPP level suggests that the possible transformations from NCNSi into NCSiN and CNNSi isomers have high kinetic barriers of 82.3 and 78.1 kcal/mol, respectively. Again, NCNSi cannot be converted directly to the CNSiN isomer. The latter is produced *via* NCSiN, requiring an extra 17.8 kcal/mol activation energy ($\Delta G^{\ddagger}$) to cross the barrier. Therefore, the occurrence of these three other isomers even at some elevated temperatures is prohibited and it is not easy to undergo this entire isomeriztion from the NCNSi compound.

Hence, we have studied this isomerization process in the presence of Ng atoms. Ng atoms are inserted into the more easily fissionable bond position of NCNSi i.e., within the NC-NSi bond (as Lewis structure of the compound is N≡C-N≡Si) to trace the other higher energy isomers during its release process. The optimized minimum energy structures and the transition states are provide in Figure 7, and the isomerization process computed at the MPW1B95/def2-TZVPP level is depicted in Figure 8.

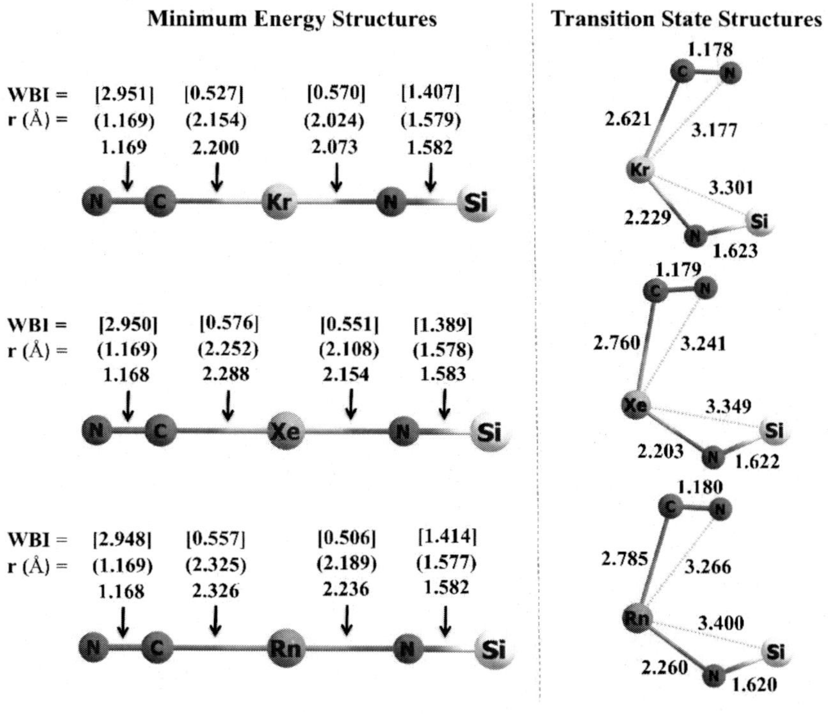

Figure 7. Minimum energy structures and TS geometries (for NCNgNSi → Ng + CNSiN dissociation) of NCNgNSi (Ng = Kr-Rn) compounds. The bond lengths are computed at the MPW1B95/def2-TZVPP (and CCSD(T)/def2-TZVPP) level. (Adapted from Pan et al. [45] with permission from John Wiley and Sons. © 2018 Wiley-VCH Verlag GmbH & Co. KGaA, Weinheim).

Table 6. The ZPE-corrected dissociation energy (D_0) and change in free energy at 298 K (ΔG) in kcal/mol for possible dissociation paths of NCNgNSi (Ng = Kr, Xe, Rn) compounds at the CCSD(T)/def2-TZVPP level. (Reprinted from Pan et al. [45] with permission from John Wiley and Sons. © 2018 Wiley-VCH Verlag GmbH & Co. KGaA, Weinheim)

Processes	D_0			ΔG		
	Kr	Xe	Rn	Kr	Xe	Rn
NCNgNSi → CN^- + Ng + NSi^+	207.0	236.1	250.2	191.0	219.9	234.1
NCNgNSi → CN^+ + Ng + NSi^-	258.2	287.4	301.5	242.3	271.2	285.3
NCNgNSi → CN + Ng + NSi	7.0	36.1	50.2	-9.8	19.1	33.2
NCNgNSi → $NCNg^+$ + NSi^-	185.0	187.7	190.2	175.9	178.4	180.8
NCNgNSi → CN^- + $NgNSi^+$	139.1	143.2	147.4	129.8	133.8	138.0
NCNgNSi → CN^+ + $NgNSi^-$	256.5	284.4	297.9	247.0	274.7	288.3
NCNgNSi → CNSiN + Ng	-66.5	-37.3	-23.2	-72.8	-43.9	-29.8
$\Delta E^{\ddagger}$ ($\Delta G^{\ddagger}$)[a]	26.4	38.1	40.2	(25.2)	(37.0)	(39.3)

[a] $\Delta E^{\ddagger}$ or $\Delta G^{\ddagger}$ values are at the MPW1B95/def2-TZVPP level.

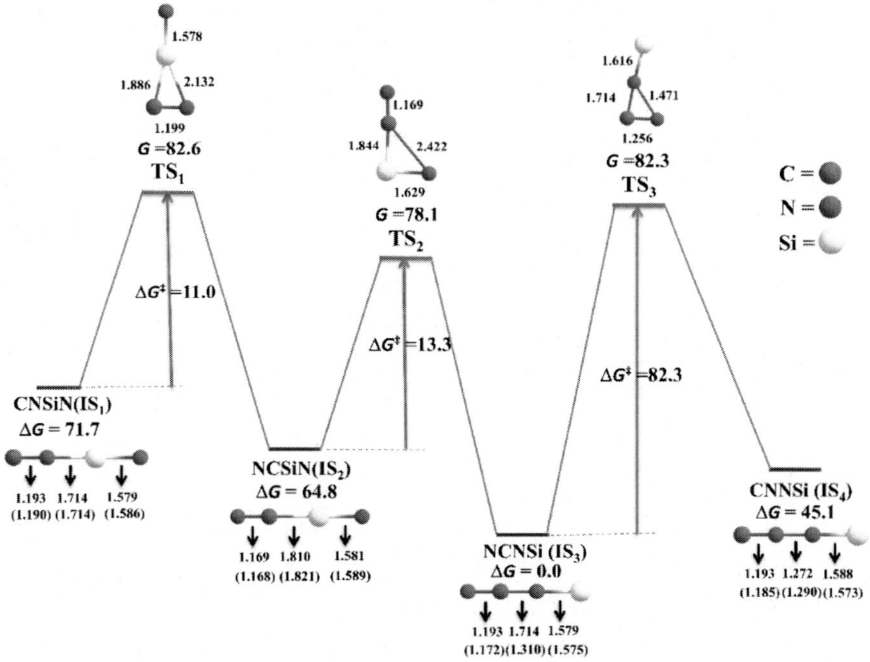

Figure 8. The overall isomerization process of NCNSi at MPW1B95/def2-TZVPP level (bond lengths within parenthesis are obtained at CCSD(T)/def2-TZVPP level). (Adapted from Pan et al. [45] with permission from John Wiley and Sons. © 2018 Wiley-VCH Verlag GmbH & Co. KGaA, Weinheim).

CCSD(T) computations on the Ng inserted compounds, NCNgNSi (Ng = Xe, Rn) reveal that they are stable *w.r.t.* possible channels of dissociation (see Table 6), and the one producing Ng and CNSiN is thermochemically feasible with appreciable kinetic protection ($\Delta G^{\ddagger}$ = 37.0 and 39.3 kcal/mol, for Xe and Rn, respectively), suggesting their possible viability at ambient temperatures. However, the Kr analogue has comparatively lower kinetic protection. The neutral Ng-release process, NCNgNSi $\rightarrow$ CNSiN + Ng, does not produce the parent compound, rather a higher energy isomer is produced in which both the NC and NSi groups flip by 180° during the dissociation. It can thus be proposed that the presence of Ng in between the NC and NSi moieties provide enough space and allow these two units to rotate during dissociation. From Table 6, it is clear that the activation free energy barrier is reduced in the presence of Ng (25.2-39.3 kcal/mol) inferring that by releasing the Ng atoms, the higher energy isomer, NCNSi, is detected.

Conclusion

This mini-review provides a thorough discussion on the isomerization in coinage metal cyanides/isocyanides, MCN/NC (M = Cu, Ag, Au) and compared the free energy barrier with the reported isomerisation of HCN and CH_3CN. From the activation free energy barrier, it is found that MNC is quite unstable both thermochemically and kinetically, apart from AuNC which exhibits a $\Delta G^{\ddagger}$ of 7.1 kcal/mol along the AuNC $\rightarrow$ AuCN isomerization (thus, might be realized at low temperatures). Also, insertion of Ng atoms within the M-C bond of MCN is investigated and the resulting MNgCN (Ng = Xe, Rn) compounds are found to be metastable as reflected by their thermochemically stable or kinetically protected dissociation channels. In the presence of Ng atoms, the corresponding $\Delta G^{\ddagger}$ values for the Cu and Ag analogues increase, which suggests that the Ng inserted isocyanide isomers, MNgNC, could be realized at low temperatures. Our investigation on the MNC/MCN isomerization in presence of different ligands also proves that ligand binding interaction increases the kinetic barrier to stabilize the MNC against its easy isomeric transformation into MCN and facilitate the trapping of transient MNC. The complete bonding situation in MNgCN is investigated and discussed here.

We have checked the validity of maximum hardness principle for the (DMI)AuNC $\rightarrow$ (DMI)AuCN isomerization and the local reactivity of

N_2AuCN/N_2AuNC species are analysed using CDFT-based reactivity descriptors.

In order to obtain higher energy isomers of NCNSi, NCNgNSi (Ng = Xe, Rn) compounds are considered and their thermochemical stability *w.r.t.* all dissociation channels at 298 K, reduced activation free energy barrier in presence of Ng, and the most important aspect i.e., the possible detection of the higher energy isomer *viz.*, CNSiN, are highlighted here.

Acknowledgments

PKC would like to thank DST, New Delhi for the J. C. Bose National Fellowship, grant number SR/S2/JCB-09/2009. RP thanks CSIR for her fellowship, and GJ thanks IIT Kharagpur and IIT Bombay.

References

[1] Kaiser Ralf, I. (2002). "Experimental investigation on the formation of carbon-bearing molecules in the interstellar medium via neutral− neutral reactions." *Chemical Reviews*, *102*, 1309–1358. doi: 10.1021/cr970004v.

[2] Bell, M. B., Feldman, P., Travers, M., McCarthy, M., Gottlieb, C. & Thaddeus, P. (1997). "Detection of HC11N in the cold dust cloud TMC-1." *The Astrophysical Journal Letters*, *483*, L61. doi: 10.1086/310732.

[3] Hoffmann, R. (1963). "An extended Hückel theory. I. Hydrocarbons". *The Journal of Chemical Physics.*, *39*, 1397–1412. doi: 10.1063/1.1734456.

[4] Van Dine, George, W. & Hoffmann R. (1968). "Isocyanide-cyanide and isoelectronic rearrangements". *Journal of the American Chemical Society*, *90*, 3227–3232.

[5] Moffat, J. B. & Tang, K. F. (1973). "The methyl isocyanide isomerization: a CNDO/2 study with partitioning of energy." *Theoretica chimica acta.*, *32*, 171–182. doi: 10.1007/BF00528489.

[6] Dewar Michael, J. S. & Kohn, M. C. (1972)."Ground states of. sigma.-bonded molecules. XVI. Rearrangement of methyl isocyanide to acetonitrile." *Journal of the American Chemical Society.*, *94*, 2704–2706.

[7] Chan, S. C., Rabinovitch, B. S., Bryant, J. T., Spicer, L. D., Fujimoto, T., Lin, Y. N. & Pavlou, S. P. (1970). "Energy transfer in thermal methyl isocyanide isomerization. Comprehensive investigation". *The Journal of Physical Chemistry*, *74*, 3160–3176.

[8] Moffat, J. (1977). "Vinyl cyanide, vinyl isocyanide, and the isomerization reaction. A theoretical study". *The Journal of Physical Chemistry.*, *81*, 82–86.

[9] Turner, B. E., Steimle, T. C. & Meerts, L. (1994). "Detection of sodium cyanide (NaCN) in IRC 10216." *The Astrophysical Journal.*, *426*, 97-100.

[10] Ziurys, L. M., Apponi, A. J., Guélin, M. & Cernicharo, J. (1995). "Detection of MgCN in IRC+ 10216: a new metal-bearing free radical". *The Astrophysical Journal.*, *445*, L47–L50.

[11] Zack, L. N., Halfen, D. T. & Ziurys, L. M. (2011). "Detection of FeCN (X 4Δ i) in IRC+ 10216: A New Interstellar Molecule". *The Astrophysical Journal Letters.*, *733*, L36. doi: 10.1088/2041-8205/733/2/L36.

[12] Guélin, M., Muller, S., Cernicharo, J., Apponi, A. J., McCarthy, M. C., Gottlieb, C. A. & Thaddeus, P. (2000). "Astronomical detection of the free radical SiCN". *Astronomy and Astrophysics.*, *363*, L9–L12.

[13] Turner, B. E., Petrie, S., Dunbar, R. C. & Langston, G. (2005). "A search for MgNC and AlNC in TMC-1: The status of metals in dark cloud cores". *The Astrophysical Journal.*, *621*, 817. doi: 10.1086/427724.

[14] Clavaguéra-Sarrio, C., Hoyau, S., Ismail, N. & Marsden, C. J. (2003). Modeling complexes of the uranyl ion UO2L2 n+: binding energies, geometries, and bonding analysis. *The Journal of Physical Chemistry A.*, *107*, 4515–4525. doi: 10.1021/jp027243t.

[15] Straka, M., Patzschke, M. & Pyykkö, P. (2003). Why are hexavalent uranium cyanides rare while U–F and U–O bonds are common and short?. *Theoretical Chemistry Accounts.*, *109*, 332–340. doi: s00214-003-0441-7.

[16] Cho, H. G. & Andrews, L. (2011). "Infrared spectra of the η2–M(NC)–CH3, CH3–MNC, and CH2= M(H)NC complexes prepared by reactions of thorium and uranium atoms with acetonitrile". *Organometallics.*, *31*, 535–544. doi: 10.1021/om200957j.

[17] Gong, Y., Cho, H. G. & Andrews, L. (2015). "Reactions of Laser-Ablated U Atoms with HCN: Infrared Spectra in Solid Argon and Quantum Chemical Calculations for HUNC". *European Journal of Inorganic Chemistry.*, *2015*, 2974–2981. doi: 10.1002/ejic.201500317.

[18] Gong, Y., Andrews, L., Liebov, B. K., Fang, Z., Garner III, E. B. & Dixon, D. A. (2015). "Reactions of laser-ablated U atoms with (CN) 2: infrared spectra and electronic structure calculations of UNC, U (NC) 2, and U (NC) 4 in solid argon". *Chemical Communications.*, *51*, 3899–3902. doi: 10.1039/C4CC09946J.

[19] Huang, Z., Sun, L., Yuan, Y., Li, Y. & Wang, X. (2016). "Theoretical insights into halogenated uranium cyanide/isocyanide compounds". *Inorganic chemistry.*, *55*, 12559–12567. doi: 10.1021/acs.inorg chem.6b01345.

[20] Rayón, V. M., Redondo, P., Valdés, H., Barrientos, C. & Largo, A. (2007). "Cyanides and isocyanides of first-row transition metals: molecular structure, bonding, and isomerization barriers". *The Journal of Physical Chemistry A., 111*, 6334–6344. doi: 10.1021/jp072434n.

[21] Grotjahn, D. B., Brewster, M. A. & Ziurys, L. M. (2002). "The first precise molecular structure of a monomeric transition metal cyanide, copper (I) cyanide". *Journal of the American Chemical Society., 124*, 5895-5901. doi: 10.1021/ja0122492.

[22] Okabayashi, T., Okabayashi, E. Y., Koto, F., Ishida, T. & Tanimoto, M. (2009). "Detection of free monomeric silver (I) and gold (I) cyanides, AgCN and AuCN: microwave spectra and molecular structure". *Journal of the American Chemical Society., 131*, 11712-11718. doi: 10.1021/ja808153g.

[23] Jana, G., Pan, S., Osorio, E., Zhao, L., Merino, G. & Chattaraj, P. K. (2018). "Cyanide–isocyanide isomerization: stability and bonding in noble gas inserted metal cyanides (metal= Cu, Ag, Au)". *Physical Chemistry Chemical Physics., 20*, 18491-18502. doi: 10.1039/C8CP02837K

[24] Bartlett, N. (1962). "Xenon hexafluoroplatinate (V) $Xe^{+}[PtF_6]^{-}$". *Proceedings of the chemical society of London*, (JUN), 218.

[25] Feldman, V. I., Sukhov, F. F., Orlov, A. Y. & Tyulpina, I. V. (2003). "Experimental evidence for the formation of HXeCCH: the first hydrocarbon with an inserted rare-gas atom". *Journal of the American Chemical Society., 125*, 4698-4699. doi: 10.1021/ja034585j.

[26] Khriachtchev, L., Tanskanen, H., Cohen, A., Gerber, R. B., Lundell, J., Pettersson, M., Kiljunen, H. & Räsänen, M. (2003). "A gate to organokrypton chemistry: HKrCCH". *Journal of the American Chemical Society., 125*, 6876-6877. doi: 10.1021/ja0355269.

[27] Khriachtchev, L., Tanskanen, H., Lundell, J., Pettersson, M., Kiljunen, H. & Räsänen, M. (2003). "Fluorine-free organoxenon chemistry: HXeCCH, HXeCC, and HXeCCXeH". *Journal of the American Chemical Society., 125*, 4696-4697. doi: 10.1021/ja034485d.

[28] Tanskanen, H., Khriachtchev, L., Lundell, J. & Räsänen, M. (2004). "Organo-noble-gas hydride compounds HKrCCH, HXeCCH, HXeCC, and HXeCCXeH: Formation mechanisms and effect of ^{13}C isotope substitution on the vibrational properties". *The Journal of chemical physics., 121*, 8291-8298. doi: 10.1063/1.1799611.

[29] Pettersson, M., Lundell, J., Khriachtchev, L. & Räsänen, M. (1998). "Neutral rare-gas containing charge-transfer molecules in solid matrices. III. HXeCN, HXeNC, and HKrCN in Kr and Xe". *The Journal of chemical physics., 109*, 618-625. doi: 10.1063/1.476599.

[30] Khriachtchev, L., Domanskaya, A., Lundell, J., Akimov, A., Räsänen, M. & Misochko, E. (2010). "Matrix-isolation and ab initio study of HNgCCF and HCCNgF molecules (Ng = Ar, Kr, and Xe)". *The Journal of Physical Chemistry A.*, *114*, 4181-4187. doi: 10.1021/jp1001622.

[31] Zhu, C., Räsänen, M. & Khriachtchev, L. (2015). "Matrix-isolation and ab initio study of HKrCCCl and HXeCCCl". *The Journal of chemical physics.*, *143*, 244319. doi: 10.1063/1.4938426.

[32] Arppe, T., Khriachtchev, L., Lignell, A., Domanskaya, A. V. & Räsänen, M. (2012). "Halogenated Xenon Cyanides ClXeCN, ClXeNC, and BrXeCN". *Inorganic chemistry.*, *51*, 4398-4402. doi: 10.1021/ic3002543.

[33] Khriachtchev, L., Lignell, A., Tanskanen, H., Lundell, J., Kiljunen, H. & Räsänen, M. (2006). "Insertion of noble gas atoms into cyanoacetylene: An ab initio and matrix isolation study". *The Journal of Physical Chemistry A.*, *110*, 11876-11885. doi: 10.1021/jp063731f.

[34] Tanskanen, H., Khriachtchev, L., Lundell, J., Kiljunen, H. & Räsänen, M. (2003). "Chemical compounds formed from diacetylene and rare-gas atoms: HKrC4H and HXeC4H". *Journal of the American Chemical Society.*, *125*, 16361-16366. doi: 10.1021/ja038610x.

[35] LeBlond, R. D. & DesMarteau, D. D. (1974). "Fluoro[imidobis(sulphuryl fluoride)]xenon. An example of a xenon–nitrogen bond". *Journal of the Chemical Society, Chemical Communications.*, 555-556. doi: 10.1039/C39740000555.

[36] Schumacher, G. A. & Schrobilgen, G. J. (1983). "Synthesis, multinuclear magnetic resonance and Raman study of nitrogen-15-enriched Xe [N (SO2F) 2] 2, an example of xenon-nitrogen bonding. Solution behavior of [15N]-F [XeN (SO2F) 2] 2+ AsF_6". *Inorganic Chemistry.*, *22*, 2178-2183.

[37] Foropoulos, Jr. J. & DesMarteau, D. D. (1982). Bis[bis(trifluoro methanesulfonyl)imido]xenon: a new compound possessing xenon-nitrogen bonds. *Journal of the American Chemical Society.*, *104*, 4260-4261.

[38] Emara, A. A. & Schrobilgen, G. J. (1987). "Fluoro (nitrile) xenon (II) cations, RC≡N-XeF+ AsF6-(R = H, CH3, CH2F, C2H5, C2F5, C3F7, or C6F5); Novel examples of xenon-nitrogen bonds and 129Xe-13C, 129Xe-1H, and 129Xe-14N nuclear spin-spin couplings". *Journal of the Chemical Society. Chemical communications.*, 1644-1646. doi: 10.1039/C398700 01644.

[39] Schulz, A. & Klapötke, T. M. (1997). "Theoretical Evidence for Two New Intermediate Xenon Species: Xenon Azide Fluoride, FXe (N3), and Xenon Isocyanate Fluoride, FXe (NCO)". *Inorganic chemistry.*, *36*, 1929-1933. doi: 10.1021/ic9613671.

[40] Fir, B., Whalen, J. M., Mercier, H. P., Dixon, D. A. & Schrobilgen, G. J. (2006). "Syntheses of [F5TeNH3][AsF6],[F5TeN (H) Xe][AsF6], and F5TeNF2 and Characterization by Multi-NMR and Raman Spectroscopy and by Electronic Structure Calculations: The X-ray Crystal Structures of α-and β-F5TeNH2,[F5TeNH3][AsF6], and [F5TeN (H) Xe][AsF6]". *Inorganic chemistry.*, *45*, 1978-1996. doi: 10.1021/ic051451t.

[41] Smith, G. L., Mercier, H. P. & Schrobilgen, G. J. (2008). "F5SN (H) Xe+; a Rare Example of Xenon Bonded to sp3-Hybridized Nitrogen; Synthesis and Structural Characterization of [F5SN (H) Xe][AsF6]". *Inorganic chemistry.*, *47*, 4173-4184. doi: 10.1021/ic702039f.

[42] Smith, G. L. & Schrobilgen, G. J. (2009). "On the Reactivity of F3S≡NXeF+: Syntheses and Structural Characterizations of [F4S≡N− Xe--N≡SF3][AsF6], a Rare Example of a N− Xe− N Linkage, and [F3S (N≡SF3) 2][AsF6]". *Inorganic Chemistry.*, *48*, 7714-7728. doi: 10.1021/ic900651n.

[43] Operti, L., Rabezzana, R., Turco, F., Borocci, S., Giordani, M. & Grandinetti, F. (2011). "Xenon–Nitrogen Chemistry: Gas-Phase Generation and Theoretical Investigation of the Xenon–Difluoronitrenium Ion F2N-Xe+". *Chemistry–A European Journal.*, *17*, 10682-10689. doi: 10.1002/chem.201101395.

[44] Bernardi, F., Cacace, F., de Petris, G., Pepi, F. & Rossi, I. (1998). "XeNO3+: A Gaseous Cation Characterized by a Remarkably Strong Xe−O Bond". *The Journal of Physical Chemistry A.*, *102*, 5831-5836. doi: 10.1021/jp980781e.

[45] Pan, S., Jana, G., Ravell, E., Zarate, X., Osorio, E., Merino, G. & Chattaraj, P. K. (2018). "Stable NCNgNSi (Ng= Kr, Xe, Rn) compounds with covalently bound C-Ng-N unit: possible isomerization of NCNSi through the release of the noble gas atom". *Chemistry – A European Journal.*, *24*, 2879-2887. doi: 10.1002/chem.201705112.

[46] Pyykkoe, P. (1995). "Predicted chemical bonds between rare gases and Au^{+}". *Journal of the American Chemical Society.*, *117*, 2067-2070. doi: 10.1021/ja00112a021.

[47] Schröder, D., Schwarz, H., Hrušák, J. & Pyykkö, P. (1998). "Cationic gold (I) complexes of xenon and of ligands containing the donor atoms oxygen, nitrogen, phosphorus, and sulphur". *Inorganic Chemistry.*, *37*, 624-632. doi: 10.1021/ic970986m.

[48] Ghanty, T. K. (2005). "Insertion of noble-gas atom (Kr and Xe) into noble-metal molecules (AuF and AuOH): Are they stable?". *The Journal of chemical physics.*, *123*, 074323. doi: 10.1063/1.2000254.

[49] Ghanty, T. K. (2006). "How strong is the interaction between a noble gas atom and a noble metal atom in the insertion compounds MNgF (M= Cu

and Ag, and Ng= Ar, Kr, and Xe)?". *The Journal of chemical physics.*, *124*, 124304. doi: 10.1063/1.2173991.
[50] Jana, G., Pan, S., Merino, G. & Chattaraj, P. K. (2017). "MNgCCH (M = Cu, Ag, Au; Ng= Xe, Rn): the first set of compounds with M–Ng–C bonding motif". *The Journal of Physical Chemistry A.*, *121*, 6491-6499. doi: 10.1021/acs.jpca.7b04993.
[51] Pal, R., Jana, G. & Chattaraj, P. K. (2020). "Ligand stabilized transient "MNC" and its influence on MNC→ MCN isomerization process: a computational study (M= Cu, Ag, and Au)". *Theoretical Chemistry Accounts*, *139*, 1-16. doi: 10.1007/s00214-019-2532-0.
[52] Zhao, Y. & Truhlar, D. G. (2004). "Hybrid meta density functional theory methods for thermochemistry, thermochemical kinetics, and noncovalent interactions: the MPW1B95 and MPWB1K models and comparative assessments for hydrogen bonding and van der Waals interactions". *The Journal of Physical Chemistry A.*, *108*, 6908-6918.
[53] Møller, C. & Plesset, M. S. (1934). "Note on an approximation treatment for many-electron systems". *Physical review.*, *46*, 618.
[54] Pople, J. A., Head-Gordon, M. & Raghavachari, K. (1987). "Quadratic configuration interaction. A general technique for determining electron correlation energies". *The Journal of chemical physics.*, *87*, 5968-5975.
[55] Dunning Jr, T. H. (1989). "Gaussian basis sets for use in correlated molecular calculations. I. The atoms boron through neon and hydrogen". *The Journal of chemical physics.*, *90*, 1007-1023.
[56] Woon, D. E. & Dunning Jr, T. H. (1994). "Gaussian basis sets for use in correlated molecular calculations. IV. Calculation of static electrical response properties". *The Journal of chemical physics.*, *100*, 2975-2988.
[57] Woon, D. E. & Dunning Jr, T. H. (1993). "Gaussian basis sets for use in correlated molecular calculations. III. The atoms aluminum through argon". *The Journal of chemical physics.*, *98*, 1358-1371.
[58] Chai, J. D. & Head-Gordon, M. (2008). "Long-range corrected hybrid density functionals with damped atom–atom dispersion corrections". *Physical Chemistry Chemical Physics.*, *10*, 6615-6620.
[59] Peterson, K. A. & Puzzarini, C. (2005). Systematically convergent basis sets for transition metals. II. Pseudopotential-based correlation consistent basis sets for the group 11 (Cu, Ag, Au) and 12 (Zn, Cd, Hg) elements. *Theoretical Chemistry Accounts.*, *114*, 283-296.
[60] Cundari, T. R. (1994). "Calculation of a methane carbon-hydrogen oxidative addition trajectory: comparison to experiment and methane activation by high-valent complexes". *Journal of the American Chemical Society.*, *116*, 340-347.

[61] Song, J. & Hall, M. B. (1993). "Theoretical studies of inorganic and organometallic reaction mechanisms. 6. Methane activation on transient cyclopentadienylcarbonylrhodium". *Organometallics.*, *12*, 3118-3126.

[62] Koga, N. & Morokuma, K. (1990.) "*Ab initio* potential energy surface and electron correlation effect in CH activation of methane by coordinatively unsaturated chlorodiphosphinerhodium (I)". *Journal of Physical Chemistry.*, *94*, 5454-5462.

[63] Reed, A. E. & Weinhold, F. (1983). "Natural bond orbital analysis of near-Hartree–Fock water dimer". *The Journal of chemical physics.*, *78*, 4066-4073.

[64] Wiberg, K. B. (1968). "Application of the pople-santry-segal CNDO method to the cyclopropylcarbinyl and cyclobutyl cation and to bicyclobutane". *Tetrahedron.*, *24*, 1083-1096.

[65] Frisch, M. J., Trucks, G. W., Schlegel, H. B., Scuseria, G. E., Robb, M. A., Cheeseman, J. R., Scalmani, G., Barone, V., Mennucci, B., Petersson, G. A., Nakatsuji, H., Caricato, M., Li, X., Hratchian, H. P., Izmaylov, A. F., Bloino, J., Zheng, G., Sonnenberg, J. L., Hada, M., Ehara, M., Toyota, K., Fukuda, R., Hasegawa, J., Ishida, M., Nakajima, T., Honda, Y., Kitao, O., Nakai, H., Vreven, T., Montgomery, Jr. J. A., Peralta, J. E., Ogliaro, F., Bearpark, M., Heyd, J. J., Brothers, E., Kudin, K. N., Staroverov, V. N., Keith, T., Kobayashi, R., Normand, J., Raghavachari, K., Rendell, A., Burant, J. C., Iyengar, S. S., Tomasi, J., Cossi, M., Rega, N., Millam, J. M., Klene, M., Knox, J. E., Cross, J. B., Bakken, V., Adamo, C., Jaramillo, J., Gomperts, R., Stratmann, R. E., Yazyev, O., Austin, A. J., Cammi, R., Pomelli, C., Ochterski, J. W., Martin, R. L., Morokuma, K., Zakrzewski, V. G., Voth, G. A., Salvador, P., Dannenberg, J. J., Dapprich, S., Daniels, A. D., Farkas, O., Foresman, J. B., Ortiz, J. V., Cioslowski, J. & Fox, D. J. (2013). *Gaussian 09, Revision D.01*. Gaussian, Inc., Wallingford CT.

[66] Bader, R. F. W. (1990). *Atoms in Molecules-A Quantum Theory*. Oxford University Press, Oxford.

[67] Huzinaga, S. & Klobukowski, M. (1993). "Well-tempered Gaussian basis sets for the calculation of matrix Hartree—Fock wavefunctions". *Chemical Physics Letters.*, *212*, 260-264.

[68] Lu, T. & Chen, F. (2012). "Multiwfn: a multifunctional wavefunction analyser". *Journal of computational chemistry.*, *33*, 580-592.

[69] Burke, K., Perdew J. & Ernzerhof, M. (1997). "Generalized Gradient Approximation Made Simple". *Physical Review Letters.*, *78*, 1396.

[70] Grimme, S., Antony, J., Ehrlich, S. & Krieg, H. (2010). "A consistent and accurate ab initio parametrization of density functional dispersion correction (DFT-D) for the 94 elements H-Pu". *The Journal of chemical physics.*, *132*, 154104.

[71] Grimme, S., Ehrlich, S. & Goerigk, L. (2011). "Effect of the damping function in dispersion corrected density functional theory". *Journal of computational chemistry.*, *32*, 1456-1465.

[72] Perdew, J. P., Burke, K. & Ernzerhof, M. (1996). Generalized gradient approximation made simple. *Physical review letters.*, *77*, 3865.

[73] Baerends, E. J., Ziegler, T., Autschbach, J., Bashford, D., Bérces, A., Bickelhaupt, F. M., Bo, C., Boerrigter, P. M., Cavallo, L., Chong, D. P., Deng, L., Dickson, R. M., Ellis, D. E., and van Faassen, M., Fan, L., Fischer, T. H., Guerra, C. F., Ghysels, A., Giammona, A., Gisbergen, S. J. A. v., Gotz, A. W., Groeneveld, J. A., Gritsenko, O. V., Gr ning, M., Gusarov, S., Harris, F. E., Hoek, P. v. d., Jacob, C. R., Jacobsen, H., Jensen, L., Kaminski, J. W., Kessel, G. v., Kootstra, F., Kovalenko, A., Krykunov, M. V., Lenthe, E. v., McCormack, D. A., Michalak, A., Mitoraj, M., Neugebauer, J., Nicu, V. P., Noodleman, L., Osinga, V. P., Patchkovskii, S., Philipsen, P. H. T., Post, D., Pye, C. C., Ravenek, W., Rodrlguez, J. I., Ros, P., Schipper, P. R. T., Schreckenbach, G., Seldenthuis, J. S., Seth, M., Snijders, J. G., Sola, M., Swart, M., Swerhone, D., Velde, G. t., Vernooijs, P., Versluis, L., Visscher, L., Visser, O., Wang, F., Wesolowski, T. A., Wezenbeek, E. M. v., Wiesenekker, G., Wolff, S. K., Woo, T. K. & Yakovlev, A. L. ADF2013.01, SCM, *Theoretical Chemistry*, Vrije Universiteit, Amsterdam, The Netherlands, 2013.

[74] Lenthe, E. V., Baerends, E. J. & Snijders, J. G. (1993). "Relativistic regular two-component Hamiltonians". *The Journal of chemical physics.*, *99*, 4597-4610.

[75] van Lenthe, E., Baerends, E. J. & Snijders, J. G. (1994). "Relativistic total energy using regular approximations". *The Journal of chemical physics.*, *101*, 9783-9792.

[76] Van Lenthe, E. V., Snijders, J. G. & Baerends, E. J. (1996). "The zero-order regular approximation for relativistic effects: The effect of spin–orbit coupling in closed shell molecules". *The Journal of chemical physics.*, *105*, 6505-6516.

[77] Van Lenthe, E., Ehlers, A. & Baerends, E. J. (1999). "Geometry optimizations in the zero order regular approximation for relativistic effects". *The Journal of chemical physics.*, *110*, 8943-8953.

[78] Parr, R. G. & Yang, W. (1984). "Density functional approach to the frontier-electron theory of chemical reactivity". *Journal of the American Chemical Society.*, *106*, 4049-4050.

[79] Morell, C., Grand, A. & Toro-Labbe, A. (2005). "New dual descriptor for chemical reactivity". *The Journal of Physical Chemistry A.*, *109*, 205-212.

[80] Morell, C., Grand, A. & Toro-Labbé, A. (2006). "Theoretical support for using the Δf (r) descriptor". *Chemical Physics Letters*. *425*, 342-346.

[81] Parr, R. G., Szentpály, L. V. & Liu, S. (1999). "Electrophilicity index". *Journal of the American Chemical Society., 121*, 1922-1924.
[82] Padmanabhan, J., Parthasarathi, R., Elango, M., Subramanian, V., Krishnamoorthy, B. S., Gutierrez-Oliva, S., Toro-Labbé, A., Roy, D. R. & Chattaraj, P. K., (2007). "Multiphilic descriptor for chemical reactivity and selectivity". *The Journal of Physical Chemistry A., 111*, 9130-9138.
[83] Parr, R. G. & Pearson, R. G. (1983). "Absolute hardness, companion parameter to absolute electronegativity". *Journal of the American chemical society., 105*, 7512-7516.
[84] Parr, R. G. & Yang, W. (1989). *Density-functional theory of atoms and molecules.* Oxford University Press, New York.
[85] Koopmans, T. (1933). "Ordering of wave functions and own energies to the individual electrons of an atom". *Physica., 1*" 104-113.
[86] Pearson, R. G. (1990). "Hard and soft acids and bases—the evolution of a chemical concept". *Coordination chemistry reviews, 100*, 403-425.
[87] Pearson, R. G. (2005). "Chemical hardness and density functional theory". *Journal of Chemical Sciences., 117*, 369-377.
[88] Chattaraj, P. K., Pérez, P., Zevallos, J. & Toro-Labbé, A. (2001). "*Ab initio* SCF and DFT studies on solvent effects on intramolecular rearrangement reactions". *The Journal of Physical Chemistry A, 105*, 4272-4283.
[89] Chamorro, E., Chattaraj, P. K. & Fuentealba, P. (2003). "Variation of the electrophilicity index along the reaction path". *The Journal of Physical Chemistry A., 107*, 7068-7072.
[90] Parthasarathi, R., Elango, M., Subramanian, V. & Chattaraj, P. K. (2005). "Variation of electrophilicity during molecular vibrations and internal rotations". *Theoretical Chemistry Accounts., 113*, 257-266.
[91] Noorizadeh, S. (2007). "Is there a minimum electrophilicity principle in chemical reactions?". *Chinese Journal of Chemistry., 25*, 1439-1444.
[92] Parr, R. G. & Chattaraj, P. K. (1991). "Principle of maximum hardness". *Journal of the American Chemical Society., 113*, 1854-1855.
[93] Chattaraj, P. K., Cedillo, A., Parr, R. G. & Arnett, E. M. (1995). "Appraisal of chemical bond making, bond breaking, and electron transfer in solution in the light of the principle of maximum hardness". *The Journal of Organic Chemistry., 60*, 4707-4714.
[94] Makov, G. (1995). "Chemical hardness in density functional theory". *The Journal of Physical Chemistry., 99*, 9337-9339.

[95] Pan, S., Solà, M. & Chattaraj, P. K. (2013). "On the validity of the maximum hardness principle and the minimum electrophilicity principle during chemical reactions". *The Journal of Physical Chemistry A.*, *117*, 1843-1852.

[96] Tonner, R. & Frenking, G. (2008). "Divalent carbon (0) chemistry, part 1: parent compounds". *Chemistry–A European Journal.*, *14*, 3260-3272.

Appendix

Table A-1. The ZPE-corrected dissociation energy (D_0, kcal mol^{-1}) and free energy change at 298 K (ΔG, kcal mol^{-1}) for different dissociation channels of LMCN calculated at the ωB97X-D/cc-pVTZ/cc-pVTZ-PP level (Reprinted from Pal et al. [51] with permission from Springer Nature. Copyright © 2019, Springer-Verlag GmbH Germany, part of Springer Nature)

$\Delta E=E(L)+E(MCN/NC)-E(LMCN/NC)$					
	D_0	ΔG		D_0	ΔG
OCCuCN	43.5	34.0	OCCuNC	49.8	40.2
H_2OCuCN	32.9	24.7	H_2OCuNC	35.1	26.8
H_2SCuCN	34.5	26.0	H_2SCuNC	38.4	29.8
N_2CuCN	27.1	18.1	N_2CuNC	31.5	22.3
NH_3CuCN	47.3	38.3	NH_3CuNC	50.9	42.4
(DMI)CuCN	74.2	65.0	(DMI)CuNC	79.7	70.5
C_2H_2CuCN	41.5	33.2	C_2H_2CuNC	47.3	38.9
C_2H_4CuCN	41.2	31.6	C_2H_4CuNC	47.4	37.6
OCAgCN	27.0	17.7	OCAgNC	30.0	20.3
H_2OAgCN	21.8	13.9	H_2OAgNC	22.0	13.9
H_2SAgCN	25.3	17.0	H_2SAgNC	27.2	18.5
N_2AgCN	14.2	5.6	N_2AgNC	15.2	6.4
NH_3AgCN	34.3	25.4	NH_3AgNC	35.6	26.4
(DMI)AgCN	60.9	51.6	(DMI)AgNC	64.5	55.0
C_2H_2AgCN	26.4	18.3	C_2H_2AgNC	28.8	20.5
C_2H_4AgCN	29.3	19.8	C_2H_4AgNC	32.3	22.5
OCAuCN	50.3	40.6	OCAuNC	64.0	54.0
	D_0	ΔG		D_0	ΔG
H2OAuCN	32.2	23.9	H2OAuNC	37.4	28.7
H2SAuCN	41.3	32.6	H2SAuNC	51.1	42.1
N2AuCN	27.7	18.4	N2AuNC	36.8	27.2
NH3AuCN	50.5	41.3	NH3AuNC	58.9	49.4
(DMI)AuCN	88.0	78.5	(DMI)AuNC	101.6	92.1
C_2H_2AuCN	45.0	36.4	C_2H_2AuNC	56.6	47.8
C_2H_4AuCN	48.9	38.9	C_2H_4AuNC	61.3	51.2

Systems	$\Delta E = E(LM) + E(CN) - E(LMCN)$		$\Delta E = E(L) + E(M) + E(CN) - E(LMCN)$		$\Delta E = E(LM^+) + E(CN^-) - E(LMCN)$		$\Delta E = E(L^-) + E(MCN^+) - E(LMCN)$	
	D_0	ΔG	D_0	ΔG	D_0	ΔG	D_0	ΔG
OCCuCN	154.5	143.5	217.5	200.4	200.6	191.2	366.1	355.7
H_2OCuCN	163.9	153.6	206.8	191.0	187.8	177.9	369.8	360.6
H_2SCuCN	153.4	143.0	208.4	192.3	186.4	177.0	369.5	360.0
N_2CuCN	154.4	144.4	201.1	184.4	200.1	190.5	360.9	350.9
NH_3CuCN	171.6	161.2	221.3	204.7	184.5	175.0	382.0	372.0
(DMI)CuCN	164.3	154.2	248.1	231.3	168.6	159.3	391.7	381.9
C_2H_2CuCN	157.9	147.9	215.5	199.5	192.2	182.8	364.0	353.7
C_2H_4CuCN	166.7	156.8	215.2	197.9	189.0	179.6	360.5	349.8
OCAgCN	133.4	121.8	133.7	116.9	178.3	168.9	346.0	335.3
H_2OAgCN	126.2	115.2	128.5	113.1	167.3	157.8	355.1	345.8
H_2SAgCN	130.8	119.6	132.0	116.2	168.2	159.0	356.7	346.9
N_2AgCN	120.5	109.4	120.9	104.9	176.4	166.9	344.4	334.4
NH_3AgCN	135.3	124.7	141.0	124.7	165.8	156.5	365.3	355.0
(DMI)AgCN	147.8	137.5	167.6	150.9	153.0	143.6	374.8	364.4
C_2H_2AgCN	132.2	119.9	133.1	117.5	170.6	160.9	345.1	334.7
C_2H_4AgCN	135.4	124.6	136.0	119.0	169.0	159.7	344.9	333.9
OCAuCN	148.2	138.0	202.4	184.9	222.1	212.4	383.0	372.3
H_2OAuCN	142.9	131.3	184.3	168.2	219.2	209.3	379.2	369.9
H_2SAuCN	145.3	134.4	193.4	176.9	205.6	195.9	386.4	376.6
N_2AuCN	140.3	128.6	179.8	162.7	228.8	218.9	371.5	361.3
NH_3AuCN	151.4	140.6	202.6	185.6	207.4	197.6	395.1	385.0
(DMI)AuCN	158.0	147.6	240.1	222.8	178.1	168.3	415.6	405.4
C_2H_2AuCN	147.9	137.3	197.1	180.7	208.4	198.7	377.4	366.9
C_2H_4AuCN	147.2	136.6	201.0	183.2	202.3	192.6	378.2	367.2

Chapter 3

Cyanide: Toxicity in the Environment

Duraisamy Udhayakumari*

Department of Chemistry, Rajalakshmi Engineering College, Chennai, India

Abstract

Cyanide (CN^-), a highly toxic anion, has drawn special interest due to its affection for many biological functions and wide applications in numerous chemical processes. To regulate cyanide in drinking water, various countries have set regulations on the maximum contaminant level for cyanide in water resources. Cyanide is one of the toxic and environmental inorganic pollutants. Due to the extreme toxicity of cyanide ions, the world health organization (WHO) recommended a tolerable limit of cyanide ion at 1.9 μM, while US environmental protection agency recommends the cyanide ion limit at 7.8 μM. Cyanide toxicity occurs because this compound strongly binds to metals, inactivating metalloenzymes such as cytochrome c-oxidase. Cyanide blocks the oxidative respiration pathway, impeding oxygen usage within tissues; the major metabolic pathway results in the formation of less toxic thiocyanate. Cyanogen chloride may be formed when cyanide polluted water is treated by chlorination. A parsimonious global cyanide cycle is proposed in this Chapter.

Keywords: cyanide, toxicity, environmental, and pollution

*Corresponding Author's E-mail: udhaya.nit89@gmail.com, udayakumari.d@rajalakshmi.edu.in

In: Cyanide: Occurrence, Applications and Toxicity
Editor: Bill M. Torres
ISBN: 978-1-68507-619-1

1. Introduction

Cyanide is considered as one of the highly toxic elements that are found both naturally and as an introduced contaminant in the environment. Cyanide is present in several compounds such as hydrogen cyanide, sodium cyanide, and potassium cyanide. According to the World Health Organization (WHO), the maximum acceptable amount of cyanide ion in drinking water is 1.9 mM. Cyanides are widely distributed among common plants in form of cyanogenetic glycosides (Baud, 2007). These glycosides hydrolyze to form hydrogen cyanide (HCN). Cyanides are in the form of gases, liquids, and solids. Hydrogen cyanide is a colorless or pale blue liquid at room temperature and volatile and forms explosive mixtures. In addition, cyanide may be dissolved into solvents to form liquids such as cyanogen chloride. Solids are typical salts made up of sodium, potassium, or calcium with the cyano group. The salts, potassium cyanide and sodium cyanide, are white powders. Cyanide vapor can ignite and explode. Hydrogen cyanide is lighter than air gas and rapidly disperses. All cyanide compounds are slowly inactivated in water (Fortin et al., 2011).

Hydrogen cyanide gas is harmful to humans and plants such as flax, lima beans, cherry, apple, peach, and apricot have characteristic high HCN potential. Hydrogen cyanide has also been used in gas-chamber executions and as a war gas. The most common forms of cyanide are free cyanide ($HCN+CN^-$), metallocyanide complexes, and organonitriles. Cyanide has a high affinity for ferric iron (Fe^{3+}), such as found in cytochrome c oxidase an important enzyme in the electron transport chain. Cyanide binds and irreversibly inhibits cytochrome c oxidase in mitochondria. This results in a shutdown of cellular respiration, soon followed by cellular death. Cyanide readily diffuses through cell walls, which is why the onset of symptoms occurs so quickly. Small amounts of cyanide are metabolized to thiocyanate through endogenous rhodanese and then excreted in the urine (Grabowska et al., 2012, Ballantyne et al., 1984, Ballantyne et al., 1984).

Cyanide ions can easily reach the environment after the usage of cyanide by various industrial activities such as gold mining, electroplating, refineries, printed circuit board manufacturing, steel, and chemical industries, synthetic fibers, resins, paper, plastics, herbicides, printing, agriculture, photography, Other cyanide sources include vehicle exhaust, releases from certain chemical industries, burning of municipal waste, and use of cyanide-containing pesticides, residential fires; polyurethane, polyacrlonitriles, organic materials, other plastics, fumigation, photographic chemicals, metallurgy, electroplating, organic synthesis, laboratory processes, dehairing of hides, manufacture of plastics, production of synthetic rubber or pesticides, fertilizers, rodent control, mass suicide or terrorist attack using cyanide, therapeutic exposure to drugs and dietary exposure to plants. Cyanide in landfills can contaminate underground water. Chlorination of water contaminated with cyanide produces the compound cyanogen chloride. Thiocyanates are a group of compounds formed from a combination of sulfur, carbon, and nitrogen and found in various foods and plants. Thiocyanides are less harmful than cyanide in humans; they are known to affect the thyroid glands, reducing the ability of the gland to produce hormones that are necessary for the normal function of the body (Banerjee et al., 2002, Botz et al., 2000, Akcil et al., 2003, Egekeze et al., 1980).

2. Common Sources of Exposure

Cyanide is readily available in an assortment of forms. Cyanide is also found in very low concentrations in foods in the form of amygdalin, a sugar compound with cyanide attached. The Compounds are widely distributed and are produced naturally by many organisms including bacteria, fungi, plants, and some insects. The well-known sources of cyanide in food are the seeds, cassava, beans, alfalfa, bamboo, cotton, corn, cherry, flex, peach, plum, potato and sorghum, lima, fruit pits from the Prunusspp like apples, bitter almonds, chokecherries, and apricots. Cyanide is used in several industries and is found at low levels in the air from car exhaust. Combustion of many substances such as nylon, plastics, wool, and silk may release hydrogen cyanide gas. Other airborne sources include emissions from chemical processing, other industries, cassava processing electroplating, metal mining processes, metal finishing and plating, metallurgy, metal cleaning, pesticide application, leather tanning, photography and photoengraving, firefighting, gas works operations, dye industries, pharmaceutical industries, and municipal waste incinerators. Smoking is another important source of cyanide. Cyanide may be found in water from discharges from organic chemical industries, iron and steelworks, and wastewater treatment facilities. Calcium cyanide is produced by the reaction of coke, coal, and limestone (Dash et al., 2009, Tran et al., 2019, Akcil et al., 2002).

3. Effects of Cyanide

3.1. Environment

Heavy metal contamination in soils and waters is a common problem encountered at many hazardous waste sites. The cyanide radical is found in a variety of naturally occurring plant compounds. These compounds include cyanogenic glycosides, glycosides, lathyrogenic compounds, indole acetonitrile, and cyanopyridine. Plants that contain cyanogenic glycosides are potentially poisonous because incomplete burning can lead to hydrolysis of the glycoside and the release of hydrogen cyanide. Cyanide is commonly used in artisanal gold mining. Gold mines are almost always set up near rivers. The excess chemicals are distributed directly into waterways, thus polluting water sources. Once it becomes embedded in soil or water, cyanide is extremely toxic to humans. The concentration of cyanide in landfills, tailings, ponds, and spills can be enough to kill the microorganisms normally responsible for their degradation. The mining industries are using cyanide to extract silver and gold from ores. Multiple chemical reactions may occur when cyanide is used in mining such as volatilization, precipitation, and oxidation (Bolstad-Johnson, et al., 2000, Borgerdinga, et al, 2005, Johnson et al., 2015). Each of these chemical reactions led to changes in toxicity based on their ability to release toxic free cyanide (Figure 1).

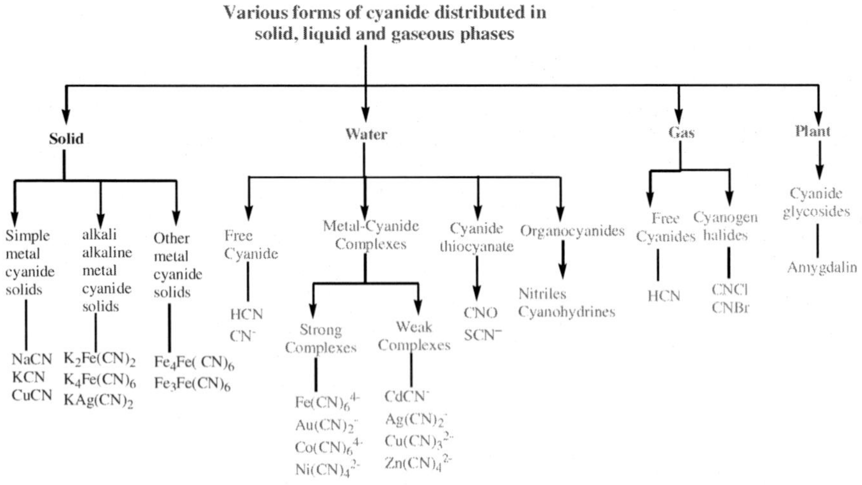

Figure 1. Various forms of cyanide distribution in the environment.

Cyanide has been the top 10% of contaminants on the priority list of hazardous contaminants since 1995. Cyanide in the environment can exist as free cyanide but typically reacts to form metal cyanide complexes. Cyanide-containing waterbodies are hazardous to wildlife, especially migratory waterfowl, bats, and accidental spills of cyanide solutions into water bodies have produced massive kills of fish and other aquatic biotas. Once unconfined into the soil matrix and waters, most heavy metals are strongly retained and their adverse effects can last for a long time. The reason for cyanide poisonings, as a consequence of food consumption, is cyanogenic glycosides in plants. The most common cyanogenic glycoside is amygdalin that can be found in seeds, pips, and kernels of fruit such as apples, peaches, almonds, cherries, plums, and apricots. Compounds containing cyanide ions are rapidly acting poison, as they disrupt the process of cellular respiration. The basic effect of cyanide activity involves combining with trivalent iron of cytochrome oxidase, which is a key enzyme of the respiratory chain (Jaszczak et al., 2017, Dursun et al., 2000, Meeussen et al., 1995, Donald et al., 2009, Kuti et al., 2006).

3.2. Animals

Classical acute cyanide poisoning happens when cyanide is bound and inhibits the ferric heme moiety form of mitochondrial cytochrome c oxidase. Cyanide stops the activity of the mitochondrial electron transport chain resulting in the arrest of aerobic metabolism, systemic hypoxia, and death from histotoxic anoxia. Cyanide also binds to other heme-containing enzymes, such as members of the cytochrome p450 family, and myoglobin. Chronic cyanide poisoning-related hypothyroidism is due to interruption of iodide uptake by the follicular thyroid cell sodium-iodide symporter by thiocyanate, a metabolite in the detoxification of cyanide. Chronic cyanide and cyanide metabolite (e.g., various glutamyl beta-cyanoalanines)-associated neuropathy toxidromes, which include diseases such as sorghum cystitis ataxia syndrome in horses, as well as various cystitis ataxia syndromes in cattle, sheep, and goats (Beelitz et al., 2017, Ilesanmi et al., 2022, Ng et al., 2019).

Zimbabwe recorded the worst ecological disaster in the previous year's which resulted in the death of over 100 elephants at Hwange National Park due to cyanide poisoning. In livestock species, the most frequent cause of acute and chronic cyanide poisoning is ingestion of plants that either constitutively contains cyanogenic glycosides or are induced to produce cyanogenic glycosides and cyanolipids as a protective response to environmental conditions. Fish are the other major group of animals and cyanide absorption probably occurs through gills and the gastrointestinal tract. In high concentrations, cyanide is toxic to aquatic life, especially fish which are one thousand times more sensitive to cyanide than humans. Most cyanide kills are usually through the accidental release of cyanides to the environment rather than inadequate procedures. In addition, the pollution from human activities, decomposing plant species which naturally occur in surface waters may contribute to the cyanide content of the water. Industrial effluents containing cyanides may find their path into waterways and increase various cyanide toxicity in aquatic organisms. Cyanide toxicity varies with animals, water pH, temperature, mineral concentration, and dissolved oxygen. Lower oxygen concentrations and an increase in temperature in water raise the toxicity of cyanide concentration (González-Valoys et al., 2022, Rice et al., 2018, Muboko et al., 2014).

3.3. Human Being

Depending on its form, cyanide may cause toxicity through inhalation, ingestion, dermal absorption, or parenteral administration. The long-term inhalation of cyanide affects the central nervous system. Short-term inhalation exposure to 100 milligrams per cubic meter (mg/m3) or more of hydrogen cyanide causes death in humans. The concentration of hydrogen cyanide of 0.3 mg/liter (270 ppm) in the air is immediately fatal to humans. Severe exposure to lower concentrations of hydrogen cyanide causes a variety of effects in humans, such as weakness, headache, nausea, increased rate of respiration, and eye and skin irritation. Humans may be exposed to cyanide as hydrogen cyanide, a cyanide salt, cyanogen, or cyanide containing compounds such as nitriles. Continuous exposure to cyanide in humans via inhalation results in effects such as headaches, dizziness, numbness, tremor, and loss of visual sharpness. Other effects include cardiovascular and respiratory effects, an enlarged thyroid gland, and irritation to the eyes and skin (Dasgupta et al., 2014). Fish retrieved from cyanide-poisoned environments, dead or alive, can probably be consumed by humans because muscle cyanide residues were considered to be lower than the currently recommended value of 50 mg/kg diet for human health protection (Figure 2).

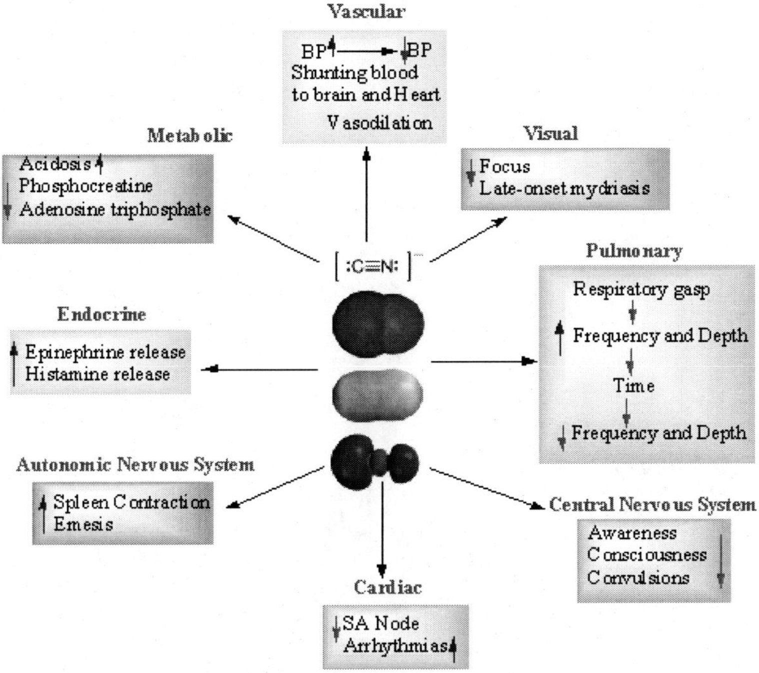

Figure 2. Cyanide effects on the human body.

The ingestion of any compound containing cyanide moiety could cause poisoning if the cyanide ion is released by digestive processes. When potassium cyanide is rapidly absorbed through the gastric mucosa, the people have become unconscious in a matter of seconds and have died within minutes. A major source of oral cyanide intake by humans is cyanogenic glycosides present in many plant foods. The appearance of symptoms is usually delayed several hours after ingestion of a cyanogenic glycoside such as that in bitter almonds or apricot kernels. The delayed absorption probably reflects the slower release of cyanide in the stomach. In inhalation metabolism, the hydrogen cyanide vapor is absorbed rapidly through the lungs. Human inhalation of 270 ppm of HCV vapor brings death immediately and 135 ppm is fatal after 30 minutes. Cyanide absorption following inhalation of very low concentrations is indicated by the observation that smokers (Tobacco smoke) have a higher concentration of thiocyanate levels in plasma and other biological fluids than for nonsmokers (Bhagavan et al., 2015). Inhalation of cyanide salts is also dangerous and the cyanide will dissolve on contact with moist mucous membranes and be absorbed into the bloodstream. Halogenated cyanogens such as cyanogen chloride or cyanogen bromide are comparable in toxicity to hydrogen cyanide when inhaled and irritate the respiratory system and pulmonary edema. Organic nitriles are another class of potential cyanide releasing compounds (Figure 3).

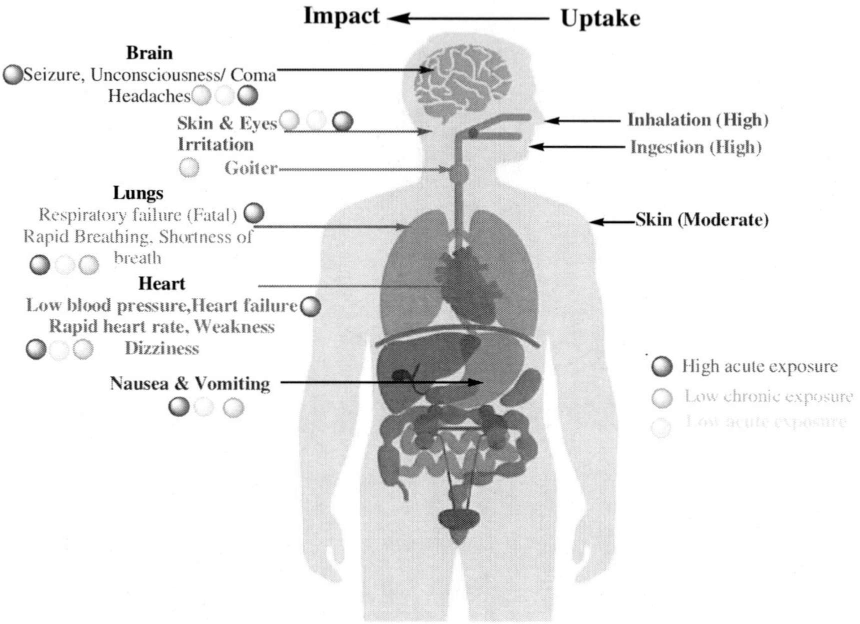

Figure 3. Major health impacts of cyanide.

Cyanide readily forms complexes with metal ions such as iron or copper. This property is probably responsible for most of its acute toxic effects. Within the bloodstream, cyanide can bind with methemoglobin to form nontoxic cyanmethemoglobin. When sodium nitroprusside is incubated with some biological materials, it releases cyanide. The most active biological material is red blood cells. Red blood cells are more active in rats than in humans, but the difference in red cell permeability to nitroprusside. Cyanide has to be released from nitroprusside via two non-enzymatic reactions: (a) a slow, nonspecific reaction with free sulfhydryl groups and (b) a fast reaction with hemoglobin. Cyanogen chloride (CNCl) should be relatively stable in blood and it decomposes slowly in an aqueous solution at up to pH=8. More than 30% of CNCl is converted to HCN in the presence of red blood cells. The lethal effects of cyanide were due to the formation of HCN and CNCl was converted to thiocyanate. Cyanide interacts with some substances in the bloodstream and most of the reactions occur intracellularly within the organs (Bhattacharya et al., 2015). The detoxification reactions, as well as biochemical reactions, have been attributed to producing the predominant toxic effects (Figure 4).

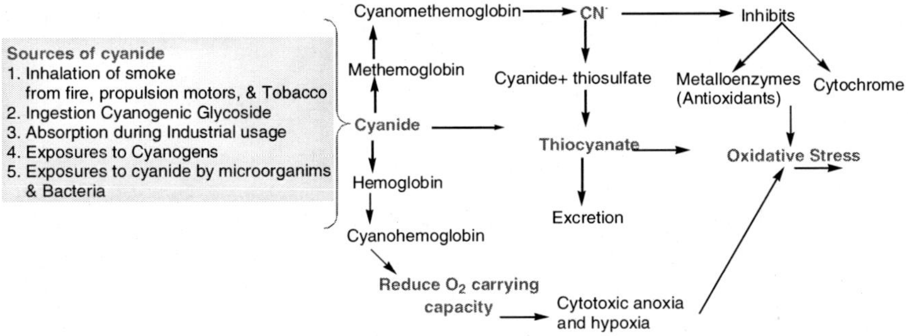

Figure 4. Different sources of cyanide exposures, its distribution, and detoxification in the body, and possible routes of male reproductive toxicity.

4. Acute Cyanide Toxicity

4.1. Inhalation Toxicity

The lethal dose of potassium cyanide was found to be 2.4 mg/kg[6]and the lethal blood level of cyanide was found to be 200–300 mg· This strength is a result of cyanide's rapid diffusion into tissues and binding to target sites. Inhalation exposure of HCN is one of the most harmful forms of CN toxicity, where the gas evades first-pass metabolism and rapidly enters the bloodstream. The inhalation of HCN along with other chemical compounds such as carbon monoxide contributes to several deaths in households and building fires. The exact contribution of HCN in fire-related deaths relative to other chemical compounds is difficult to assess because of the breakdown of CN in the blood postmortem Intravenous and inhalation exposures to cyanide produce the most rapid onset of signs, within seconds to minutes, and toxicity from ingestion of cyanide or cyanogenic compounds can occur within minutes to hours. Cyanide has a strong "knockdown effect" as a result of high concentrations of HCN a fire victim could lose consciousness and consequently prevent an escape, and therefore die from carbon monoxide poisoning or carbon monoxide and HCN. Acute inhalation toxicity revelation in animals is always challenging. The effect of the gas always depends on the concentration and the duration of exposure. Inhalation of HCN is distinctive in that death may be delayed as a result of differences in tidal volume, respiration rate, and time, which dictates the total concentration of CN inhaled during an exposure (Allen et al., 2015). The animals are exposed to diverse doses of HCN displayed typical acute toxic signs such as convulsions, tachycardia, ataxic movements, and respiratory depression (Figure 5).

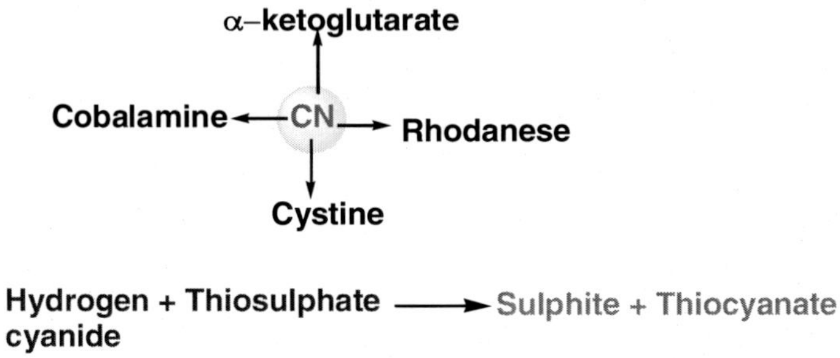

Figure 5. Detoxification of HCN molecules by livings.

Cyanides are present in low concentrations in the environment; therefore several intrinsic biochemical pathways for cyanide detoxification exist. Acute human HCN exposure leads to a chain of effects to include the altered sense of smell, tachypnea, dyspnea, nausea, ataxia, unconsciousness, palpitations, convulsions, and asphyxiation. The most important route for cyanide excretion is through the formation of thiocyanate, which is subsequently excreted in the urine through the kidneys. Thiocyanate is formed primarily in the liver, directly through the activity of the rhodanese enzyme and indirectly via the enzymes 3-mercaptopyruvate sulfurtransferase and thiosulfate reductase. These enzymes are responsible for combining cyanide and sulfur to form thiocyanate. Despite these intrinsic mechanisms for cyanide detoxification, these enzyme systems are easily overwhelmed during cyanide poisonings because of the body's limited supply of sulfur. In humans who ingest 4.6–15 mg/kg as KCN, the symptoms like coma, reduced information processing, enlarged heart, nausea, vomiting, Parkinsonian-like symptoms, decreased verbal fluency, hyperventilation, inaudible heart sounds, albuminuria, and generalized muscular rigidity are observed and oral cyanide has led death within minutes after ingestion of cyanide (Peddy et al., 2006).

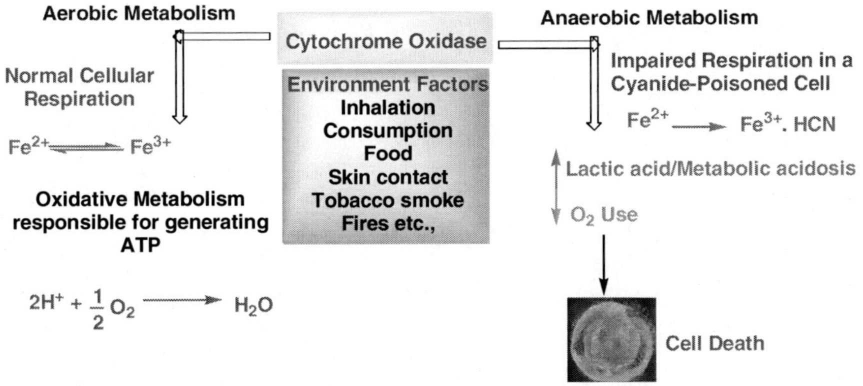

Figure 6. Cyanide inactivates mitochondrial cytochrome oxidase to inhibit cellular respiration.

4.2. Dermal Toxicity

Dermal toxicity is the ability of a substance to poison people or animals by contact with the skin. Due to accidental exposure, the dermal exposure of toxicity is occurring. More than 42% of workers exposed to 15 ppm HCN developed rashes. As per the Obiriet al. study the human health risk assessment from exposure to free cyanide via dermal contact of surface/underground water by resident adults close to mining companies with wastewater effluent and found risks for acute exposure very high. In this society, many of the residents attributed most of the unknown causes of death to dermal contact with cyanide water and accidental ingestion. Cyanide in solution is absorbed across intact skin because of its lipid solubility. Depending on the cyanide composition Species differences can pose an issue and give different results. Other factors also affect the rate of dermal absorption such as follicle concentration, skin hydration, and occlusion of skin, the thickness of stratum corneum, the lipid content of the skin, adnexal structures, and physiochemical properties of cyanide (Obiri et al., 2006).

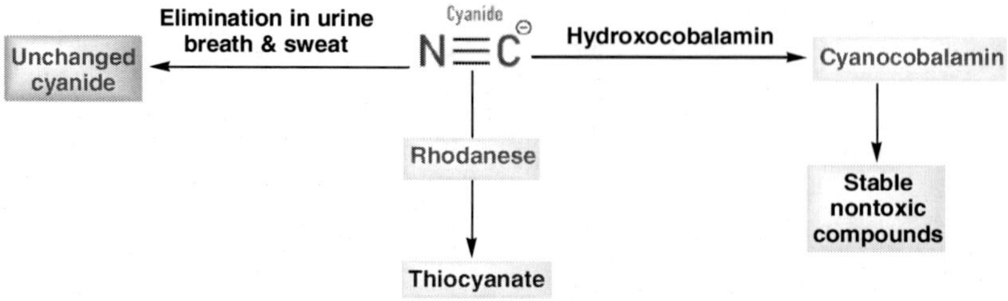

Figure 7. Detoxifying and eliminating cyanide.

Cyanide binds to cytochrome oxidase in the mitochondria, resulting in inhibition of the electron transport chain and loss of aerobic metabolism and generation of adenosine triphosphate. Nitrites react with oxyhemoglobin to form methemoglobin, which draws cyanide out of the mitochondria and forms cyanomethemoglobin. The enzyme rhodanese combines thiosulfate with cyanomethemoglobin to form thiocyanate, which is excreted by the kidneys and eliminated in the urine. Methemoglobin is converted back to oxyhemoglobin by the enzyme methemoglobin reductase (Figure 6). The primary endogenous mechanism of cyanide detoxification is metabolism in the liver by rhodanese to thiocyanate, which is a nontoxic compound excreted in the urine. Minor routes of metabolism include binding of cyanide to hydroxocobalamin (vitamin B_{12}) to form cyanocobalamin, which is excreted in urine; and oxidation of cyanide to thiocyanate and other stable nontoxic compounds through various enzymatic and nonenzymatic pathways (Rao et al., 2013). These endogenous mechanisms can detoxify only very small amounts of cyanide—0.017 mg of cyanide per kg of body weight per minute in the average person (Figure 7).

5. Chronic Cyanide Exposure

5.1. Human Disease

Combustion of carbon and nitrogen-containing products such as polyurethane, silk, wool, melamine resins, polyacrylonitriles, and synthetic rubber can liberate cyanide. Cyanide poisoning is dramatic, acute, and severe in the classic presentation, although subacute and chronic exposures may also produce adverse effects. Exposure to a lower level of cyanide over a long period results in increased blood cyanide levels, which can result in weakness and a variety of symptoms, including permanent paralysis, nervous lesions, hypothyroidism, and miscarriages. Health manifestations such as congenital malformations, malnutrition, myelopathy, and neurological disorders have been attributed to chronic cyanide toxicity. Chronic cyanide poisoning can occur if you're exposed to 20 to 40 ppm trusted source of hydrogen cyanide over a substantial period. Chronic human cyanide poisoning may result from a wide range of exposures and may present clinical materializations in cases of repeated low-dose cyanide exposure. The central nervous system is highly dependent on aerobic metabolism; chronic cyanide exposure can produce neuronal and axonal deterioration that is evident in Parkinson's. Cyanide binding to cytochrome c oxidase and disrupts aerobic cellular respiration. The mechanism of cyanide poisoning is the production of a histotoxic anoxia by binding the CN^- ion to the ferric iron (Fe^{3+}) of mitochondrial cytochrome oxidase. This results in inhibition of oxidative phosphorylation, anaerobic metabolism, lactic acid accumulation, and decreased ATP production. Chronic low-dose exposure to cyanide has been associated with several disease circumstances, particularly those of demyelinating nervous conditions. While demyelinating syndromes have been described after acute cyanide poisoning, they have been generalized anoxic laceration and are not specific for cyanide poisoning (El Ghawabi et al., 1975).

5.2. Dietary Health Hazards

Large amounts of cyanogenic plants like cassava are consumed, an increased incidence of optic atrophy and Nigerian nutritional ataxic neuropathy. Chronic exposure to cyanogenic plants in foodstuffs, especially in tropical areas where such root staples as cassava are consumed in great quantities, have been associated with disorders of cyanocobalamin metabolism, and with such various neurological diseases as amblyopia, optic atrophy, and neuropathies. Dietary deficiencies of vitamin B_{12} and the presence of achlorhydria are also associated with these neuropathies and confuse the aetiological relationship of cyanogens to their development. A frequent source of chronic cyanide ingestion is through the diet foods such as cassava root, staples, and lima beans. Free cyanide in food and drink is usually derived from the hydrolysis of cyanogenic glycosides and includes linamarin in cassava, dhurrin in sorghum species as well as amygdalin and prunasin in almonds, cherry laurel, and apricot pits (Figure 8). Tobacco amblyopia, Leber's hereditary optic atrophy, Subacute combined degeneration of the cord, Tropical ataxic neuropathy, Lathyrism (animals), Possible teratogenicity (animals), and Goitre have been associated with increased plasma thiocyanate levels and postulated as being due to low-dose chronic cyanide exposure (Newton et al., 1981).

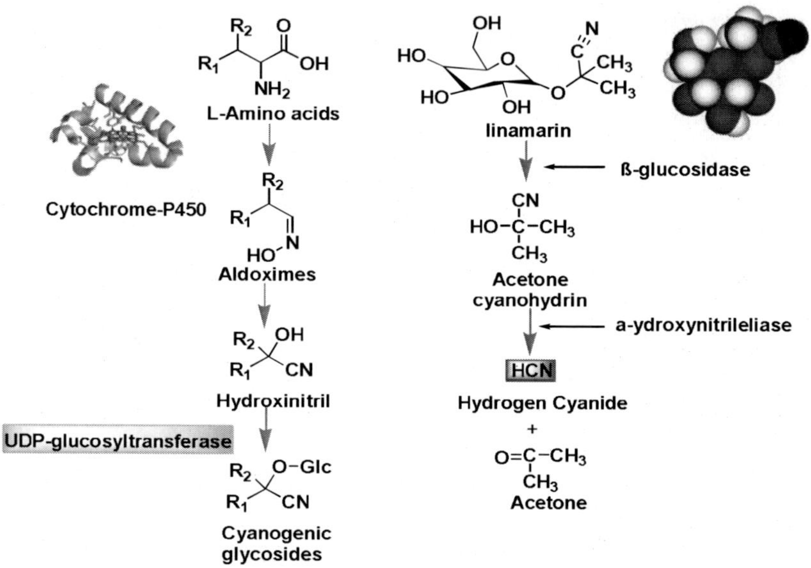

Figure 8. Common cyanogenicglucosides.

5.3. Cassava Consumption

Cassava, known also as tapioca, manioc, and yuca, originated in South America but is the major nutritional food. In most of the regions, cassava is making it an attractive food source and can endure extreme drought. Agricultural crises precipitated by drought, war, or surges in cassava trade can lead to food shortages and necessitate that families again rely on insufficiently processed cassava, thereby increasing the incidence of Tropical ataxic neuropathy (TAN). If cassava is not processed properly cyanide removal is not complete and consumption may be dangerous. The amount of cyanide in these foods is usually below acutely lethal levels for humans, although in uncooked cassava, the levels can be dangerously high. The disadvantages of using cassava as a food source include the potential cyanide toxicity of both the roots and leaves and the low protein content of the roots. Bitter cassava has a higher amount of glucosides. Drought and poor soil increase the level of cyanogenicglucosides. Linamarin is the main cyanogenicglucoside in cassava. When plant tissue is mechanically disintegrated or fermented, the glucoside is broken down into cyanohydrins by β-glucosidase in the plant cell and cyanohydrins are metabolized to, or spontaneously decompose into, HCN. Careful preparation of the cassava root, such as fermentation or soaking in water for three to five days, reduces this cyanogenic potential to levels safe for consumption. β-cyanoalanine synthase converts cyanide released from the cyanogen linamarin to cyanoalanine plus hydrogen sulfide in the presence of cysteine. Cyanoalanine is converted by a nitilase to asparagine, which is then converted to aspartate and ammonia by asparaginase. In this way, cyanide nitrogen can be incorporated into the free amino acid pool of the plant (Hall, et al., 1992, Madhusudanan et al., 2008). An alternative pathway can potentially result in cyanide detoxified to thiocyanate by rhodanese (Figure 9).

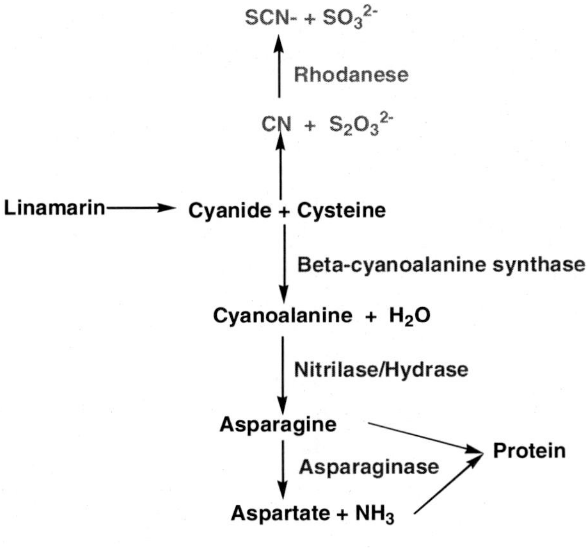

Figure 9. Cyanogen metabolism in cassava roots.

5.4. Cyanide Toxicity from Amygdalin

Amygdalin is a naturally occurring chemical compound found in many plants, most notably in the seeds (kernels) of apricots, bitter almonds, apples, peaches, cherries, lemons, nectarine, and plums. Amygdalin was isolated by French chemists in 1830 and it was first used as a treatment for cancer in Russia in

1845. In the United States, amygdalin was used in the 1920s. In the 1980s, The National Cancer Institute-sponsored Phase 1-2 clinical trials but found no evidence to support the clinical benefits of amygdalin in cancer treatment. Amygdalin was associated with cyanide poisoning, especially from oral ingestion and the Food Drug Administration (FDA) banned the sale of amygdalin as a medicinal product. Amygdalin is classified as a cyanogenic glycoside because each amygdalin molecule includes a nitrile group, which can be released as the toxic cyanide anion by the action of a beta-glucosidase (Figure 10). Eating amygdalin will cause it to release cyanide in the human body, and may lead to cyanide poisoning. Upon ingestion, amygdalin is hydrolyzed to cyanide by beta-glucuronidase in the small intestine. Oral intake of 500 mg of amygdalin may contain as much as 30 mg of cyanide. Amygdalin produces neurologic damage in humans similar to that seen in persons suffering from tropical ataxic neuropathy, a disorder attributed to chronic cyanide exposure. Oral amygdalin is estimated to be 40 times more potent than intravenous form due to its enzymatic conversion to hydrogen cyanide in the gastrointestinal tract. Severe cases may present with cyanosis, coma, convulsions, cardiac arrhythmias, cardiac arrest, and death. Thiocyanates were used extensively for the treatment of hypertension but were abandoned because of their toxic side effects which included the occurrence of goiter. In contrast to thiocyanate poisoning, chronic cyanide intoxication has received little attention. Chronic cyanide poisoning (i.e., long-term exposure to sub-lethal concentrations of cyanide) can be more insidious, with headaches, weakness, chest and abdominal pain, itch, and rash (Dang et al., 2017).

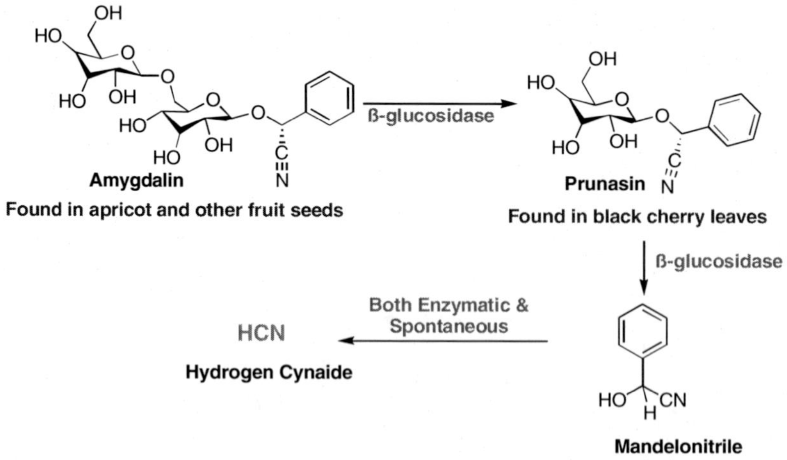

Figure 10. Hydrolysis of amygdalin and prunasin to produce cyanide.

5.5. Thyroid Dysfunction

Iodine is required for the normal function of the thyroid gland, which transports iodide into the thyroid through the sodium-iodide symporter and incorporates iodide in the thyroid hormone molecules. Insufficient delivery of iodide to the thyroid gland triggers compensatory hyperplasia of the gland referred to as a goiter. Exposure to thiocyanate corresponds to a decrease in iodine intake. Cessation of smoking, reduction of industrial pollution, and improved diet will reduce the role of thiocyanate in thyroid disease. Large amounts of thiocyanate are generated in people with a high intake of cyanide from tobacco smoking, from cyanide in food, or industrial pollution of the environment with cyanide. In individuals exposed to high levels of thiocyanate, adverse effects may be prevented by an increase in iodine intake. In areas of low iodine intake, thiocyanate exposure increases the risk of developmental and other iodine deficiency disorders. The effect of thiocyanate is to hamper the utilization of iodide; the main effect of thiocyanate is to worsen iodine deficiency. By this mechanism, thiocyanate is one of the most important environmental compounds influencing the occurrence of thyroid disease. The thiocyanate ion is similar to the iodide ion in size and charge, thiocyanate competitively inhibits iodine uptake by the thyroid gland, and the iodine is excreted in the urine. Thiocyanate is a competitive inhibitor of the sodium iodide symporter at thiocyanate levels normally found in the blood. With high ingestion of inefficiently prepared cassava, the potential for chronic cyanide overload exists, which in turn creates a high level of serum thiocyanate. This results in a decreased absorption of iodide by the thyroid and a subsequent increase in iodide excretion into the urine. Neurologic abnormalities and severe mental retardation occur in cases of severe iodine deficiency and may be exacerbated at high enough thiocyanate concentrations. The dominance of goiter and cretinism was significantly correlated with serum thiocyanate concentration and urinary iodide excretion (Chandra et al., 1980).

5.6. Cyanogenic Glycosides

Certain plants contain glycosides with a cyanide group. When attacked by pests, these plants release cyanide that results in the formation of hydrogen cyanide, which kills the invaders. Some of the important cyanogenic glycosides include amygdalin, sambunigrin, vicianin, lanimarin, and lotaustralin. Examples of some edible plants with cyanogenic glycosides include almond, apricot, sorghumcherry, plum, lima beans, cassava, prunasin, stone fruit, and bamboo shoots. In humans, acute poisoning with cyanide has been reported with overconsumption of preparations made from bitter almonds and cyanide-rich apple seeds. The cyanide interferes with cellular respiration and may cause seizures, loss of consciousness, and even cardiac arrest if taken in sufficiently large amounts. The enzyme rhodanase detoxifies cyanide to produce thiocyanate. Thethiocyanategenerated during detoxification of cyanide in the body is believed to interfere with iodine uptake because of its structural similarity. It increases the iodine requirement in iodine-deficient populations. The thiocyanate thus acts as an additional factor for the development of goiter in such populations. It is a neurologic condition associated with the intake of improperly processed cassava fruitThecyanogenic glycosides of the plant, namely linamarin and lotaustralin, are released by the plant to defend it against animals. These cyanide compounds produce konzo that is characterized by a sudden onset of spastic parapresis. Konzo has been reported in the African continent. Tropical ataxic neuropathy (TAN) is another condition that is associated with chronic consumption of products derived from cassava and consequent exposure to cyanide. It manifests in the form of neurologic symptoms that mainly affect vision and hearing abilities (Vetter, 2017).

Chronic cyanide exposure may directly cause dietary deficiencies. Tobacco amblyopia is a vitamin B_{12} deficiency and is characterized by vision loss, which is often restored after smoking cessation. Varying by brand, tobacco cigarettes can contain as much as 150 μg HCN in each cigarette. Chronic exposure to cyanide in cigarette smoke produces derangements of cobalamin concentrations in humans. Urinary thiocyanate excretion and urinary vitamin B_{12} excretion were increased and serum B_{12} levels were lower in smokers compared to non-smokers. The urinary thiocyanate excretion was significantly and negatively correlated with serum vitamin B_{12} concentrations. Cyanocobalamin, a form of vitamin B_{12} inactivated by chelation of cyanide, was detected in the plasma of heavy cigarette and pipe tobacco smokers. Cyanide effects in pregnancy, fetal development, delivery, and fertility are long termed (Downey et al., 2015).

6. Cyanide Poisoning by Smoke Inhalation

Smoke is a mixture of heated air, gases, fumes, aerosols, vapors, and solid and liquid particles in suspension. Smoke toxicity depends on the composition of the fuel, the availability of oxygen, the completeness of combustion, and the heat intensity. Even within the same building, the smoke composition can vary considerably with the duration of the fire and location. Toxic combustion products (TCPs) from naturally occurring materials and synthetic polymers in the modern environment contribute to smoke toxicity. Smoke inhalation is a common cause of cyanide poisoning during fires, resulting in injury and even death. In many cases of smoke inhalation, cyanide has increasingly been recognized as a significant toxicant. The diagnosis of cyanide poisoning remains very difficult, and failure to recognize it may result from inadequate or inappropriate treatment. Cyanide may evolve in large quantities from the combustion of nitrogen-containing materials found in modem plastic furnishings and some natural fabrics such as wool and silk. The expansion of the polymer industry makes it likely that hydrogen cyanide will be encountered with increasing frequency in fires. Few major factors can cause death in fires (Table 1). Inhalation of smoke in enclosed-space fires is the most common etiology of acute cyanide poisoning. Eight major factors such as direct consumption by the fire, very high temperatures, oxygen deficiency, and presence of carbon monoxide, presence of other toxic gases, presence of smoke, development of fear, shock, and panic, and secondary fire effects due to mechanical reasons can cause death in fires (Jones et al., 1987).

Table 1. Toxic combustion products

Product of combustion	Common sources
Acids/Aldehydes	Cellulose acetate, polyvinyl acetate, cotton, paper, wood
Ammonia	Nylon, silk, wood, and resins
Carbon monoxide and dioxide	All organic materials
Cyanide and hydrogen cyanide	Nylon, paper, resins, acrylonitriles, polyurethane, silk, and wool
Isocyanates	Polyurethane
Nitrogen oxides	Celluloid, cellulose nitrate fabrics paper, wood, and petroleum products
Sulfur oxides and hydrogen sulfide	Meat, petroleum products, wood, paper, rubber, and hair

CN inhibits mitochondrial cytochrome C oxidase, a component of the respiratory chain, thus inhibiting oxidative phosphorylation and causing tissue hypoxia because of the inability to use the delivered O_2. CN poisoning affects the central nervous system, respiratory system, and cardiovascular system, proportionate to the concentration of cyanide inhaled. The symptoms of CN toxicity vary from tachycardia, tachypnea, dyspnea, drowsiness, and headache at low concentration to cardiac arrhythmias, hypotension, convulsions, paralysis, cardiorespiratory collapse, and coma at high concentration (Table 2). Blood cyanide levels can be measured to confirm exposure to CN (Lee et al., 2007). Cyanide impairs cellular utilization of oxygen, which can result in cellular hypoxia and a variety of cardiac manifestations. As per the air monitoring study, cyanide is present in fires around 50%. Ambient air monitoring for CO and CN^- can now be done on the fireground in real-time which can alert firefighters to the potential for poisoning. A significant cyanide poisoning component should be suspected in smoke inhalation victims with otherwise unexplained respiratory failure or a persistent anion-gap metabolic acidosis (Lee-Chiong et al., 1999). Inhalation of polymer pyrolysis products including hydrogen cyanide in smoke has been shown to result in cardiotoxicity with elevated creatine phosphokinase (CPK) activity and an increased number of heartbeats. The production of hydrogen cyanide during combustion and pyrolysis is both material- and temperature-dependent. Cyanide is released from nitrogen-containing material and relatively high temperatures are required. As per the reported study ambient atmospheric concentrations of cyanide and other gases at structural fires, hydrogen cyanide was found in only 12% of studied fires and the maximum measured concentration was 40 ppm (Lowry et al., 1985).

Hydrogen cyanide and carbon monoxide in fire smoke are at least additive toxicants and may indeed be Synergistic. Higher yields of hydrogen cyanide occur in small oxygen vitiated flaming fires in closed compartments and are fully developed, post-flashover fires in open compartments at high temperatures. In such fires, the systemic asphyxiant effects of carbon monoxide, hydrogen cyanide, and low ambient oxygen levels together with dense, irritant smoke which impedes escape attempts are the greatest hazards. As per the report, 80% of fire victims die from smoke inhalation rather than burns. The combination of carbon monoxide and cyanide resulted in incapacitation and death. Smoke inhalation is also a major factor in non-trauma-related deaths in aircraft crashes and cyanide plays a detrimental role in fatalities from aviation accident fires. The ability of oxygen to access the cytochrome oxidase enzyme is essential to normal, life-sustaining cellular respiration. When cyanide reaches the cell, it binds with cytochrome a3, a subunit of the cytochrome oxidase complex and blocking the oxygen and preventing the cell from being able to use it (Grabowska et al., 2012, Eckstein et al., 2006, Baud et al., 1991).

Table 2. Smoke inhalation – associated cyanide poisoning

Low-inhaled concentrations	Moderate to high-inhaled concentrations
Faintness	Prostration
Flushing	Tremors
Anxiety	Cardiac arrhythmia
Excitement	Convulsions
Perspiration	Stupor
Vertigo	Paralysis
Headache	Coma
Drowsiness	Respiratory depression
Tachypnea	Respiratory arrest
Dyspnea	Cardiovascular collapse

Conclusion

Cyanide compounds are used in several industrial processes, including mining, metallurgy, manufacturing, and photography, due to their ability to form stable complexes with a range of metals. Cyanide has been employed extensively in electroplating, in which a solid metal object is immersed in a plating bath containing a solution of another metal with which it is to be coated to improve the durability, electrical resistance, and/or conductivity of the object. Cyanide toxicity affects all body systems, producing profound effects on the heart and brain as toxicity progresses. Progression and severity of toxicity vary according to route, concentration, and duration of exposure. Use in industrial processes is the main origin of cyanide in the environment, but cyanide is also released from biomass burning, volcanoes, and natural biogenic processes from higher plants, bacteria, and fungi. The cyanogenic compounds, which are converted to cyanide in the body, naturally occur in many plant foods, including cassava root, almonds, millet sprouts, lima beans, soy, spinach, bamboo shoots, and sorghum. Exposure to cyanide also occurs from smoking. Thiocyanate (SCN^-), the primary metabolite of cyanide, is found in plasma or blood at approximately 0.5–4 μg/L in nonsmokers and approximately 6–22 μg/L in smokers.

Cyanide exerts its acute effects, including CNS depression, convulsions, coma, and death, by binding with cytochrome c oxidase, a key enzyme in the production of ATP by way of oxidative phosphorylation. The steep dose-response occurring with acute high-dose exposures is thought to be due to cyanide overload, resulting in saturation of detoxification pathways that metabolize cyanide to less acutely toxic intermediate compounds. At lower dose rates, an efficient detoxification system (primarily via rhodanese with sulfur donors as the rate-limiting factor) catalyzes the transformation of cyanide to thiocyanate, its primary metabolite. Thiocyanate is not known to be acutely toxic, although long-term exposures can adversely affect the thyroid gland via iodide uptake inhibition and decreased thyroid hormone synthesis. Chronic cyanide exposure can occur through several routes. Poorly regulated industry and insufficiently processed crops high in cyanogenic glycosides are particular risks for long-term exposure to cyanide. Inadequate detoxification can result in serious morbidity and mortality, including motor control deficits and paralysis as seen in konzo and tropical ataxic neuropathy, loss of vision, endocrine dysfunction, and death. Properly processing cassava has been shown to reduce the cyanogenic content of the roots to safe levels, even in cultivars known to be rich in cyanogenic glycosides. Personal protective equipment has been developed to help prevent occupational exposure to cyanide. Point-of-care detection of cyanide exposure and next-generation cyanide medical countermeasures is needed to rapidly identify and treat cyanide exposure victims, especially in cases of acute cyanide exposure.

References

Akcil, A., Mudder, T., (2003). Microbial destruction of cyanide wastes in gold mining: Process review. *Biotechnol. Lett*, 25, 445-450.

Akcil., (2002). The first application of cyanidation process in Turkish gold mining and its environmental impacts. *Miner. Eng*, 15, 695-699.

Allen, A.R., Booker, L., Rockwood, G. A., (2015). *Toxicology of Cyanides and Cyanogens: Experimental, Applied and Clinical Aspects,* Editors: Alan H. Hall, Gary E. Isom and Gary A. Rockwood. 1-20.

Ballantyne, B. (1984a). Comparative acute toxicity of hydrogen cyanide and its salts. In R. E. Lindstrom (ed.) *Proceedings of the Fourth Annual Chemical Defense Bioscience Review.* US Army Medical Research Institute of Chemical Defense, Maryland: Elsevier.

Ballantyne, B. (1984b). Relative toxicity of carbon monoxide and hydrogen cyanide in combined atmospheres. *Toxicologist*, 4, 69.

Banerjee, A.R., Sharma., Banerjee, U.C., (2002). The nitrile-degrading enzymes: current status and future prospects. *Appl. Microbiol. Biotechnol*. 60, 33–44.

Baud, F. J., (2007). Cyanide: Critical issues in diagnosis and treatment. *Hum. Exp. Toxicol*, 26, 191–201.

Baud, F. J., Barriot, P., Tuffis, V., et al. (1991). Elevated blood cyanide concentrations in victims of smoke inhalation. *N Engl J Med*, 325, 1761–6.

Beelitz, J., Kill, C., Feldmann, C., Wulf, H., Vogt, N., Veit, F., Dersch, W., (2017). Toxic inhalation with hydrogen cyanide prompts a significant increase of mean pulmonal arterial pressure in an animal model. *Resuscitation*, 118, e104.

Bhagavan, N. V., Ha, C. E., (2015). Hemoglobin, *Essentials of Medical Biochemistry*, 489-509.

Bhattacharya, R., Flora, S. J. S., (2015). *Handbook of Toxicology of Chemical Warfare Agents, Cyanide Toxicity and its Treatment,* Editor(s): Ramesh C. Gupta, 301-314.

Bolstad-Johnson, D. M., Burgess, J.L., Crutchfield, C.D., Storment, S., Gerkin, R., Wilson, J.R., (2000). Characterization of firefighter exposures during fire overhaul. *AIHAJ* 61,636–641.

Borgerdinga, M., Klusb, H., (2005). Analysis of complex mixtures: cigarette smoke. *Exp. Toxicol. Pathol*, 57,43–73.

Botz, M., Mudder, T., (2000). Modeling of natural cyanide attenuation in tailings impoundments. *Mining, Metallurgy & Exploration*, 17, 228-233.

Chandra, H., Gupta, B.N., Bhargava, S.K., Clerk, S.H., Mahendra, P.N., (1980). Chronic Cyanide Exposure — A Biochemical and Industrial Hygiene Study. *J. Anal. Toxicol*, 4, 161–165.

Dang, T., Nguyen, C., Tran, P. N., (2017). *Physician Beware: Severe Cyanide Toxicity from Amygdalin Tablets Ingestion.* Case Reports in Emergency Medicine; New York, Article ID. 4289527, 3 pages.

Dasgupta, A., Wahed, A., (2014). Cyanide Poisoning: Common Poisonings Including Heavy Metal Poisoning, *Clinical Chemistry, Immunology and Laboratory Quality* Control. Editor(s): Amitava Dasgupta, Amer Wahed, 337-351.

Dash, R. R., Gaur, A., Balomajumder, C., (2009). Cyanide in industrial wastewaters and its removal: A review on biotreatment. *J. Haz. Mat*, 163, 1-11.

Donald, G.B., (2009). Cyanogenic foods (cassava fruit kernels and cycadseeds). *Medical Toxicology of Natural Substances*, 55,336–352.

Downey, J.D., Basi, K.A., DeFreytas, M.R., Rockwood, G. A., (2015). *Toxicology of Cyanides and Cyanogens: Experimental, Applied and Clinical Aspects*: Chronic cyanide exposure Case studies and animal models, Editors: Alan H. Hall, Gary E. Isom and Gary A. Rockwood, 21-40.

Dursun, A.Y., Aksu, Z., (2000). Biodegradation kinetics of ferrous (II) cyanide complex ions by immobilized *Pseudomonas fluorescens* in a packed bed column reactor. *Process Biochem* 35, 615–622.

Eckstein, M. and Maniscalco, P. M. (2006). Focus on smoke inhalation – the most common cause of acute cyanide poisoning. *Prehosp Disaster Med*, 21, 49–55.

Egekeze, J.O., Oehme, F. W., (1980). Cyanides and their toxicity: A literature review. *Vet. Q*, 2, 104-114.

El Ghawabi,S. H., Gaafar,M.A., El-Saharti, A. A., Ahmed, S.H., Malash,K.K., Fares, R., (1975). Chronic Cyanide Exposure: A Clinical, Radioisotope, and Laboratory Study. *Br. J. Ind. Med*, 32, 215-219.

Fortin, J. L., Judic-Peureux, V., Desmettre, T., (2011). Hydrogen cyanide poisoning in a prison environment: A case report. *J. Correct. Health. Care*, 17, 29–33.

González-Valoys, A. C., Arrocha, J., Monteza-Destro, T., Vargas-Lombardo, M., MaríaEsbrí, J., Garcia-Ordiales, E., Jiménez-Ballesta, R., JesúsGarcía-Navarro, F., Higueras, P., (2022). *J. Environ. Manage*, 302, 113979.

Grabowska, T., Skowronek, R., Nowicka, J. &Sybirska, H. (2012). Prevalence of hydrogen cyanide and carboxyhaemoglobin in victims of smoke inhalation during enclosed-space fires: a combined toxicological risk. *Clin.Tox*, 50, 759–763.

Hall, A.H., Rumack, B.H. (1992). *Clinical and Experimental Toxicology of Organophosphates and Carbamates*: 42 - Incidence, presentation and therapeutic attitudes to anticholinesterase poisoning in the USA, Editor(s): Bryan Ballantyne, Timothy C. Marrs, 471-481.

Ilesanmi, O. B., Akinmoladun, A. C., Elusiyan, C. A., Ogungbe, I. V., Olugbade, T. A., Olaleye, M.T., (2022). Neuroprotective flavonoids of the leaf of AntiarisafricanaEnglea against cyanide toxicity. *J. Ethnopharmacol*, 282, 114592.

Jaszczak, E., Polkowska, Z., Narkowicz, S., Namieśnik., (2017). Cyanides in the environment—analysis—problems and challenges. *Environ. Sci. Pollut. Res.* 24, 15929-15948.

Johnson, C.A., (2015). The fate of cyanide in leach wastes at gold mines: an environmental perspective. *Appl. Geochem*, 57,194–205.

Jones, J., Jo McMullen, M., Dougherty, J., (1987). Toxic Smoke Inhalation: Cyanide Poisoning in Fire Victims, *Am. J. Emerg. Med*, 317–321.

Kuti, J.O., Konoru, H.B., (2006). Cyanogenic glycosides concent in two edibles leaves of tree spinach (*Cnidoscous* spp.). *J. Food. Compos. Anal*, 19, 556–561.

Lee, J., Mukai, D., Kreuter, K., Mahon, S., Tromberg, B., Brenner, M., (2007). Potential interference by hydroxocobalamin on cooximetry hemoglobin measurements during cyanide and smoke inhalation treatments. *Ann. Emerg. Med*.802–805.

Lee-Chiong, T. L., (1999). Smoke inhalation injury: When to suspect and how to treat. *Postgrad Med*, 105, 55–62.

Lowry, W. T., Juarez, L., Petty, C. S., and Robert, B. (1985). Studies of toxic gas production during actual structural fires in the Dallas area. *J Forens Sci*, 30, 59–72.

Madhusudanan, M., Menon, M.K., Ummer, K., Radhakrishnanan, K., (2008). Clinical and Etiological Profile of Tropical Ataxic Neuropathy in Kerala, South India. *EurNeurol*, 60, 21–26.

Meeussen, J.C.L., Van Riemsdijk, W.H., Van der Zee, S.E.A.T.M. (1995). Transport of complexed cyanide in soil. *Geoderma*, 67, 73–85.

Muboko, N., Murindagomo, F., (2014). Wildlife control, access and utilization: lessons from legislation, policy evolution and implementation in Zimbabwe. *J. Nat. Conserv*, 22, 206–211.

Newton, G.W., Schmidt, E.S., Lewis, J.P., Conn E., Lawrence R., (1981). Amygdalin toxicity studies in rats predict chronic cyanide poisoning in humans. *West. J. Med*, 134, 97-103.

Ng, P.C., Hendry-Hofer, T. B., Witeof, A. E., Mahon, S. B., Brenner, M., Boss, G. R., Bebarta, V.S., (2019). Efficacy of Oral Administration of Sodium Thiosulfate and Glycine in a Large, Swine Model of Oral Cyanide Toxicity. *Ann Emerg Med*, 74, 423-429.

Obiri, S., Dodoo, D., Okai-Sam, F., and Essumang, D., (2006). Non-cancer health risk assessment from exposure to cyanide by resident adults from the mining operations of Bogoso Gold Limited in Ghana. *Environ Monit Assess*, 118, 51–63.

Peddy, S. B., Rigby, M. R., and Shaffner, D. H. (2006). Acute cyanide poisoning. *Pediatr Crit Care Med*, 7, 79–82.

Rao, P., Singh, P., Yadav, S., Gujar, N., and Bhattacharya, R., (2013). Acute toxicity of some synthetic cyanogens in rats: Time-dependent cyanide generation and cytochrome oxidase inhibition in soft tissues after sub-lethal oral intoxication. *Food Chem Toxicol*, 59, 595–609.

Rice, N.C., Rauscher, N.A., Langston, J.L., Myers, T.M., (2018). Behavioral toxicity of sodium cyanide following oral ingestion in rats: Dose-dependent onset, severity, survival, and recovery. *Food Chem. Toxicol*, 114, 145-154.

Tran, Q.B., Lohitnavy, M., Phenrat, T., (2019). Assessing potential hydrogen cyanide exposure from cyanide-contaminated mine tailing management practices in Thailand's gold mining. *J. Environ. Manage*, 249, 109357.

Vetter, J., (2017). Plant Cyanogenic Glycosides. In: Gopalakrishnakone, P., Carlini C., Ligabue-Braun, R. (eds.) *Plant Toxins. Toxinology.* Springer, Dordrecht. *Plant Toxins*. 287-317.

Chapter 4

Cyanide Occurrence and Treatment in the Gold Mining Industry

Caroline Dale* **and Henri Jogand**
Biotechnologies Expert, BS&P - Scientific & Technological Expertise Department, Veolia

Abstract

Cyanide leaching is the most commonly used method for gold extraction. Although cyanide recovery and destruction is undertaken, excess water still contains cyanide residual and cyanide compounds such as cyanate and thiocyanate. These compounds are toxic to aquatic life and need to be removed from water before discharge to the environment to comply with local regulations.

Commercially available methods for cyanide removal will be described including reverse osmosis, chemical oxidation and biological treatment. A focus on biological treatment using Moving Bed Biofilm Reactor (MBBR) technology will be made. Extensive testing at a mine site in Ghana has demonstrated that biological treatment is a cost effective and sustainable solution for cyanide removal from mine effluents.

* Corresponding Author's E-mail: caroline.dale@veolia.com.

In: Cyanide: Occurrence, Applications and Toxicity
Editor: Bill M. Torres
ISBN: 978-1-68507-619-1

Gold Leaching Process

It is estimated that gold represents around 0.005 part per million of the Earth's crust. It is often found in its elemental state in igneous rocks. Gold is present at low concentrations in ores making aqueous chemical extraction the only economically viable method of extraction. As one of the noble metals, gold is not soluble in water, extraction typically involves a leaching step in which metallic gold is oxidized and dissolved in an alkaline solution. A complexant, such as sodium cyanide, with which gold forms a stable complex and an oxidant, such as atmospheric oxygen is required.

Gold used to be extracted using mercury, however cyanidation, developed in 1887, has long been the preferred process for leaching gold from ores because of its high gold recoveries, robustness and relatively low cost compared to mercury. Cyanidation allowed profitable extraction of lower grade ores (Verbrugge et al., 2021). It still accounts for the majority of gold production today, consuming around 13% of world cyanide usage.

Ore containing gold is crushed and finely ground to form a slurry. A froth flotation step may be included to concentrate low grade sulfide ores. Heavier solids sink and are discharged to the TSF while sulphides and moderately coarse gold floats are mixed with a dilute solution of sodium cyanide (0.01 to 0.05% concentration) in the presence of atmospheric oxygen. Under these conditions, the cyanide anions will dissolve and combine with the gold cations in the ore to form a metal complex.

The gold solubilization process is characterized by the following reaction, known as the 'Elsner Equation' as shown in Equation 1.

$$4\ Au(s) + 8\ NaCN(aq) + O_2(g) + 2H_2O(l) \rightarrow 4\ Na[Au(CN)_2](aq) + 4\ NaOH(aq) \quad (1)$$

Cyanide ions bond with gold present in the ore, forming a soluble complex which can be easily separated from the other insoluble metals, a process known as leaching. Stoichiometrically, 0.26kg of CN^- is required to extract 1 kg of gold. In practice, the amount of cyanide required will depend on the type of ore and its metal content such as Fe and Cu as cyanide will also form complexes with these metals. Gold can be separated from the "pregnant solution" bearing gold–cyanide complexes by reduction with Zn, either using Zn shavings or zinc dust. Zn has a higher affinity for cyanide than gold hence adding elemental zinc powder in the absence of oxygen reduces gold ions to its free metal form (Kuyucak et al. 2013). Gold can then be recovered by filtration.

A schematic diagram of the gold extraction process is shown in Figure 1.

Cyanide residues from the leaching process were generally discharged to large ponds called tailings storage facilities (TSF) without any treatment. The discharge of untreated cyanide residue to tailings ponds has resulted in death of wildlife, particularly birds using the ponds for drinking. Failure of TSF dams have resulted in a number of high profile spills of contaminated effluent, causing significant damage to the environment. The Romanian Aural Gold spill, caused by a tailings pond dam failure in 2000 resulted in severe contamination of the Tisza and Danube Rivers. Following this environmental incident, the International Cyanide Management Code, known as the cyanide code was instigated. The code, which is voluntary, aims to improve the management of cyanide used in gold and silver mining and to improve the protection of human health and the reduction of environmental impacts (International Cyanide Management Institute, 2002). The code provides guidelines for cyanide management but also includes a framework for auditing mines by third parties to ensure that the guidelines are put into practice.

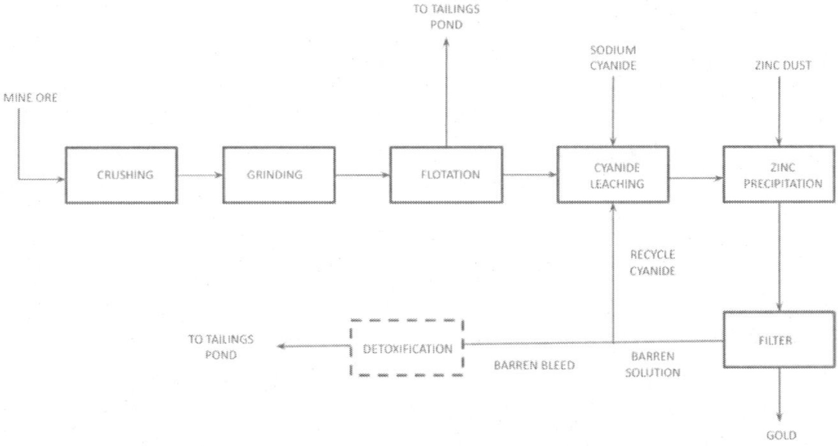

Figure 1. Gold extraction process schematic.

While there has been research aimed at identifying and developing less toxic leaching agents, none has yet made any significant inroad into the dominant position of cyanide as the reagent of choice at the vast majority of gold mines worldwide (Hilson, 2006). The use of cyanide allows for cost effective extraction of gold from low grade ore, making its replacement a challenge.

To reduce costs and because of cyanide toxicity, recycling of cyanide bearing solutions has been introduced at some mines. Mines with a positive water balance will need to discharge excess water to the TSF regardless of whether cyanide recycling is implemented or not. When recycling is not possible and where excess water needs to be discharged, the cyanide solution will generally be "detoxified" (the cyanide will be destroyed) before it is discharged to the tailings pond (Botz, 2001).

Detoxification of Barren Bleed Effluent

Treatment to lower cyanide levels entering tailings ponds is mandatory in most regions of the world today to reduce the impact on wildlife. A limit of 50 mg/L Weak Acid Dissociable (WAD) cyanide is deemed safe to wildlife and accepted as "best practice" amongst the gold mining industry.

The most common detoxification process for the barren bleed effluent before discharge to the TSF is the INCO Sulfur Dioxide/Air process which was developed and patented by the International Nickel Company in 1984.

This process uses SO_2 in the presence of air and of a copper catalyst to oxidize cyanide to cyanate:

$$SO_2 + O_2 + H_2O + CN^- \rightarrow OCN^- + SO_4^{-2} + 2\,H^+ \qquad (2)$$

SO_2 is added as a mixture of liquid SO_2, sodium sulfite (Na_2SO_3) and sodium metabisulfite ($Na_2S_2O_5$). The theoretical SO_2 requirement is 2.64 kg/kg CN^-. Actual dosages tend to be 3 - 5 g/g for clear barren solutions and 4 - 7 g/g for tailings slurries (EPA 1984). The reaction is typically carried out at a pH of about 8.0 to 10.0 in one or more agitated tank and lime is added to neutralize the acid formed in the reaction. Typical lime usage ranges between 3.0 to 5.0 g/g of CN^- oxidized.

Metals previously complexed with cyanide, such as copper, nickel and zinc, are precipitated as metal-hydroxide compounds - the process generates a slurry that needs to be disposed of.

A major disadvantage of the INCO process is that it increases the sodium and sulfate concentrations making the control of conductivity, sulfate and sodium concentration more challenging. Conventional concentration processes such as RO downstream of the INCO process may quickly become incompatible with an effective salt and water management strategy at the mine. Over time, discharge of RO brine to TSF combined with TSF effluent recycling can cause a salt saturation of the TSF which can be costly to treat.

Alternate processes exist where the oxidizer is hydrogen peroxide or ozone. The end product (cyanates) is the same.

$$CN^- + H_2O_2 \rightarrow OCN^- + H_2O \quad (3)$$

The limitations of hydrogen peroxide treatment include handling and costs. Hydrogen peroxide is a hazardous material and will require the installation of specific equipment which will increase the capital costs of the plant. One advantage of using hydrogen peroxide is that it will not increase the Total Dissolved Solids (TDS) concentration of the effluent unlike the INCO process.

If the detoxification process is operated properly, cyanides are present at low concentrations in a gold mine effluent, typically 1 to 5 mg/L of free cyanide (Mudder et al., 2001), levels at which the water acute toxicity for birds and mammals is not an issue and thus it can be discharged and stored in the tailings pond. In water scarce areas, TSF effluent can be recycled into the process. In this case, a solids removal step and cyanide removal polishing step is required.

Other compounds which may result in toxicity are also found in gold extraction process residues. Copper is present in gold mine effluents whether it is present in the ore or not, as it is an essential catalyst required in the detoxification processes.

Thiocyanates are another major constituent of gold mine effluent. Thiocyanates are a by-product of the reaction of free cyanide in the gold leaching process with sulfide ions present in the ore, typically as pyrite (Habashi, 1967).

$$S^{-2} + CN^- + \frac{1}{2}O_2 \rightarrow SCN^- + 2\ OH^- \quad (4)$$

Ammonia is another significant component of gold mine effluents. Ammonia is present both as explosive residue when the common ammonium nitrate fuel oil explosive is used (Bailey, 2011) and because of the hydrolysis of the cyanate ions

$$OCN^- + 2H_2O \rightarrow HCO_3^- + NH_3 \quad (5)$$

Natural Degradation in the TSF

Natural degradation of free cyanide in the tailings pond occurs via a range of processes including volatilization, photodecomposition, chemical oxidation, microbial oxidation, chemical precipitation, hydrolysis and precipitation on solids. A schematic of the processes involved is shown in Figure 2.

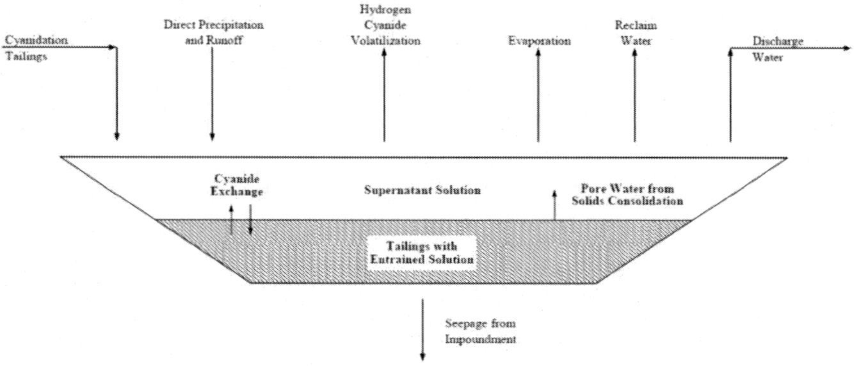

Source: Botz 2016.

Figure 2. Schematic of natural cyanide degradation in tailings ponds.

The degree of natural degradation of free cyanide in the tailings pond will vary depending on climatic conditions such as temperature and sun intensity (Kuyucak et al. 2013). The reduction in free cyanide may be limited during winter conditions or in wet weather when the retention time in the tailings pond is reduced. Turbidity will also have a significant impact on the penetration of UV through the water phase hence will impact the degree of natural cyanide degradation in the TSF. Residual solids from the extraction process frequently have poor settleability; the efficiency of natural degradation is thus frequently limited.

To ensure that water can be continuously discharged to the environment from the tailings pond in compliance with discharge limitations, mines may be required to install dedicated cyanide treatment technologies downstream of the TSF. Treatment methods may be chemical or biological in nature. Chemical processes may consist of alkaline chloride oxidation (alkaline chlorination), hydrogen peroxide oxidation, Inco SO_2/Air oxidation, Hemlo/Golden Giant (copper and iron sulphate) precipitation and Acidification–Volatilization–Regeneration (AVR). Chemical methods of removing or eliminating cyanide are widely reported (Mudder et al., 2001, Kuyucat et al., 2013) and will not be detailed in this chapter.

Cyanide Treatment Solutions for Environmental Discharge

The acute toxicity threshold of free cyanide to aquatic life is significantly lower than the toxicity to wildlife such as birds and mammals; values ranging from 0.05 to 0.16 mg/L have been reported (EPA 1976). Cyanide treatment is thus often required downstream of the tailings pond before discharge to the natural environment. Discharge limits imposed will vary depending on geographic location; a typical discharge limit for free cyanide would be 0.2 mg/L.

Chemical Oxidation Using Ozone

Cyanide oxidation with ozone is a two-step reaction: first cyanide is oxidized to cyanate, with ozone reduced to oxygen per Equation 6:

$$CN^- + O_3 \rightarrow CNO^- + O_2 \quad (6)$$

Then cyanate is hydrolyzed, in the presence of excess ozone, to bicarbonate and nitrogen and oxygen per Equation 7:

$$2\ CNO^- + 3O_3 + H_2O \rightarrow 2\ HCO_3^- + N_2 + 3O_2 \quad (7)$$

Ozone requirements range between 1.8 and 2 g O_3 per g CN^- for the first step reaction; the ozone can be produced on site. Ozone is relatively expensive to produce and this has limited its use for cyanide destruction, particularly for large water flows, but may find application in small-volume polishing applications (Mudder et al., 2001). Ozone does have a number of advantages; notably the complete oxidation of cyanide and cyanide compounds such as cyanate and thiocyanate to harmless nitrogen gas as well as a relatively simple installation in an existing treatment line.

Chemical Oxidation Using Sodium Metabisulfite (SMBS)

As an alternative to the INCO sulphur dioxide/air process, sodium metabisulfite (SMBS), $Na_2S_2O_5$, is frequently used to convert free cyanide to less toxic cyanate according to Equation 8.

$$CN^- + Na_2S_2O_5 + O_2 + 2\,H_2O \rightarrow CNO^- + H_2SO_4 + Na_2SO_3 \quad (8)$$

The stoichiometric requirement for SMBS is 7.3 g/g CN^-, however in practice, the dosage is significantly higher. Chemical addition is made in continuously stirred tanks. Concentration in free CN^- of less than 1 mg/L can be achieved with this system however the dosage required can be as high as 400% of the stoichiometric requirements (Hewitt, Aranguri et al., 2018).

Reverse Osmosis

Cyanide can be separated from water using membranes with reverse osmosis. In reverse osmosis, pressure is applied to a cyanide solution needing treatment, water is forced through a membrane impermeable to cyanide. Cyanide, amongst other soluble compounds such as ammonia and nitrate, is concentrated in a brine solution which is usually recycled to the tailings pond. An accumulation of salts is thus observed over time. As mentioned earlier, the combination of TSF effluent recycling and RO can result in salt saturation of the tailings pond; requiring expensive treatment to maintain effluent recycling without affecting gold recovery.

Due to the cost of reverse osmosis and the pre-treatment required upstream to avoid membrane fouling, reverse osmosis is generally not applied solely for cyanide removal but more generally for mine effluent treatment to meet stringent discharge limits.

Biological Treatment

Biological treatment utilizes a range of microorganisms such as bacteria, fungi, yeast or even algae to degrade compounds present in wastewater. By providing the right environment such as dissolved oxygen, pH and temperature, specific organisms can develop. In cyanide removal applications, biological treatment offers an alternative to chemical treatment with the added benefit that free cyanide is fully converted to innocuous nitrogen gas which is released to the atmosphere if a complete nitrogen removal (nitrification - denitrification) plant is installed.

The biological pathways of cyanide degradation are well described in literature (Naveen et al., 2011; Ibrahim et al., 2015 and Alvillo- Riviera et al., 2021, Ebbs 2004) and can either consist of hydrolysis, oxidation, reduction, and/or substitution/transfer. The operating conditions such as pH, dissolved oxygen availability will influence which biological pathway dominates. The reactions involved are detailed in Table 1 below.

Table 1. Cyanide degradation

Under **aerobic** conditions, the following reactions occur :

(1) conversion of cyanide to cyanate:

$$CN^- + \frac{1}{2}O_2 \rightarrow OCN^-$$

(2) conversion of cyanate to ammonia-nitrogen:

$$OCN^- + 3H_2O \rightarrow NH_4^+ + HCO_3^- + OH^-$$

(3) conversion of ammonia-nitrogen to nitrate:

$$NH_4^+ + \frac{3}{2}O_2 \rightarrow NO_2^- + 2H^+ + H_2O$$

$$NO_2^- + \frac{1}{2}O_2 \rightarrow NO_3^-$$

Under **anoxic** conditions, the following reaction occurs :

$$NO_3^- + \text{Organic Matter} \rightarrow N_2 + CO_2 + H_2O$$

A wide range of naturally occurring organisms can detoxify free cyanide; the degradation potentials of some microorganisms on free and complex cyanides and thiocyanate have been reported (Ibrahim et al., 2015). It should be noted that the rate at which cyanide complexes are degraded will depend on the stability of the complex.

There are two main types of biological treatment: suspended growth and fixed growth systems. In suspended growth, biomass develops as flocs within the reactor; these flocs are separated from the water stream in a downstream clarifier and recycled to the reactor to maintain the biomass inventory necessary for treatment. Fixed growth systems utilize a support for biofilm growth; the support can either be fixed or mobile. Examples of fixed growth systems are rotating biological contactors (RBC), trickling filters and moving bed biofilm reactors (MBBR).

In mining applications, fixed growth systems are more frequent as large flows need to be treated - fixed growth systems allow to decouple the biomass retention time from the hydraulic retention time without the need for large recirculation flows. Fixed growth systems are also more suited to changes in loadings and are more resistant to toxic compounds.

The Homestake mine in the US used a two-stage RBC system to reduce free cyanide from 4.1 mg/L down to 0.06 mg/L from a blend of mine water and tailings pond effluent. The RBCs were operated between 1984 to the mine closure in 2002 (Maier et al., 2009). Blending relatively warm mine water with tailings pond water to maintain the temperature of the combined wastewaters at a sufficiently elevated temperature year round allowed biological activity to be sustained through the seasons.

The Moving Bed Biofilm Reactor (MBBR) developed in the 1990's is particularly suitable for nitrogen removal in challenging applications such as mining. The MBBR utilizes a polyethylene carrier, AnoxK™5, with a large protected surface area for biofilm development as shown in Figure 3.

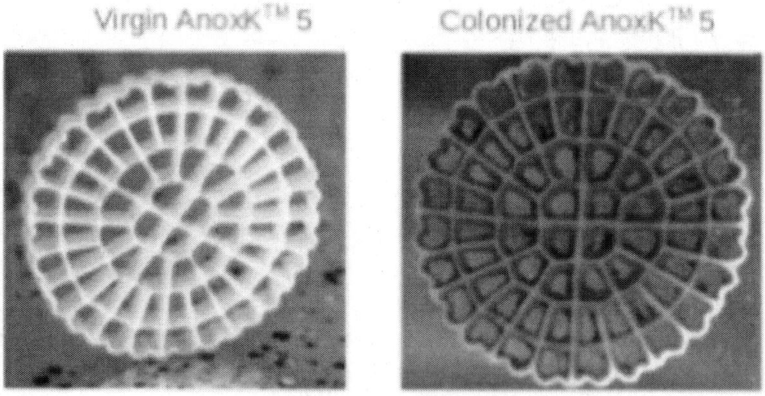

Figure 3. Virgin AnoxK™ 5 and colonized AnoxK™ 5.

The carriers have a density close to that of water, making it possible to keep the carriers in suspension and constant movement by either aeration or mechanical mixing, thus allowing operation under aerobic or anoxic mode. The carriers are retained within the tank using sieves. A schematic diagram of an aerobic and an anoxic MBBR is shown in Figure 4.

Bacteria such as slow growing cyanide oxidizing bacteria and nitrifying bacteria develop on the carriers without the need for long retention times, resulting in compact reactors. Installing several reactors in series allows the development of specific microbial communities in the individual reactors, thereby optimizing the operating conditions for specific pollutant removal and allowing high removal rates to be achieved. Other compounds present in mining effluent such as ammonia, nitrate, cyanate and thiocyanate are also removed by biological treatment however WAD cyanide is a known toxic to nitrifying bacteria. The impact of free cyanide concentration between 0.1 and 1 mg/L was demonstrated on a nitrifying activated sludge as shown in Figure 5; concentration of 0.2 mg/L severely impacted the nitrification rate whilst at 1 mg/L free cyanide nitrification was completely inhibited (Kim et al., 2008). Installation of a dedicated cyanide removal MBBR upstream of a nitrification MBBR will increase the nitrification rate in the nitrification MBBR, thereby allowing a more compact system to be installed.

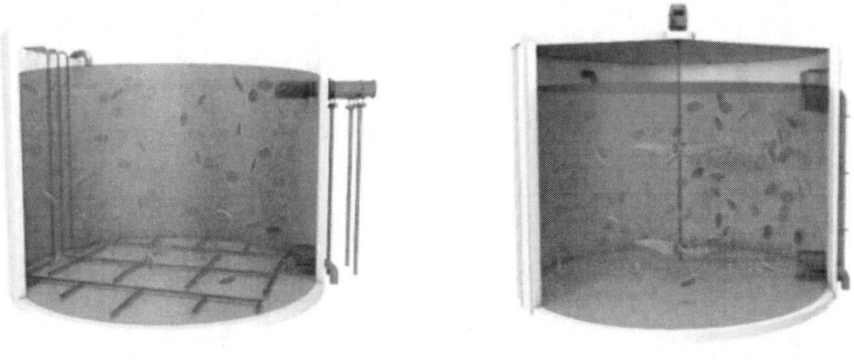

Figure 4. Aerobic MBBR with aeration grid and Anoxic MBBR with mechanical mixer.

In recent years, the use of MBBR for mine water treatment has grown; cyanate and thiocyanate removal using MBBR has been demonstrated at full scale (Villemur et al., 2015; Kwofie et al., 2021).

Extensive testing on cyanide removal has been undertaken at a gold mine using a cyanide extraction process. Effluent from the cyanide extraction process is discharged to a TSF. The TSF effluent is then pumped to a clarifier for solids removal. Clarified effluent is either recycled to the extraction process or discharged to the environment.

The characteristics of the clarified tailings pond effluent are presented in Table 2.

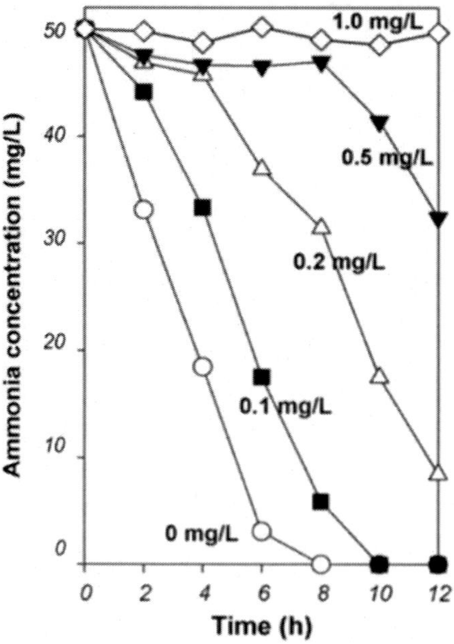

Source: Kim et al., 2008.

Figure 5. Impact of free cyanide concentration on ammonia removal in activated sludge.

Table 2. Characteristics of the clarified tailings pond effluent

		Min	Average	95%ile
pH		6.73	7.73	8.41
EC	µS/cm	953	1442	1733
Turbidity	NTU	0.78	24	50
SCOD	mg/L	4	61	117
TSS	mg/L	1	18	42
SO4	mg/L	99	248	327
Cu	mg/L	0.1	0.3	0.6
NH3-N	mg/L	0.01	3.8	9.4
NO2-N	mg/L	25.2	44.5	56.1
NO3-N	mg/L	18.8	36.1	48.4
CN free	mg/L	0.002	0.3	1.10

Factors such as turbidity, retention time in the tailings pond, dilution from rain and temperature will affect the natural degradation of cyanide in the tailings pond, resulting in fluctuations in free cyanide concentration in the clarified TSF effluent. The variability in free cyanide concentration in the clarified tailings pond effluent is shown in Figure 6.

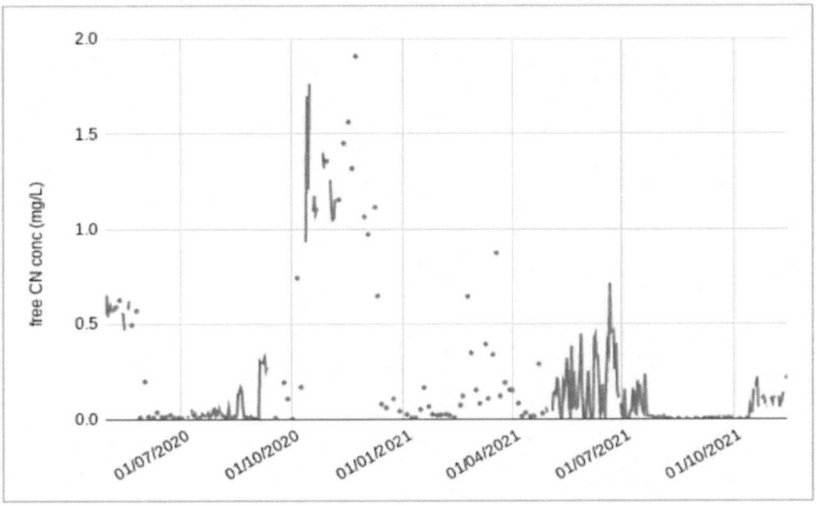

Figure 6. Variation in the clarified tailings pond effluent.

The free CN concentration is frequently higher than the discharge limit of 0.2 mg/L imposed by the local regulations and thus requires treatment before release to the environment. In this particular case, the mine can recycle non-compliant effluent to other storage facilities and recycle the water within the process due to the negative water balance at the site. Discharge to the environment is not required on a continuous basis. When effluent needs to be discharged, post treatment is required depending on the quality of the clarified TSF effluent. SMBS dosing is used in the event of excess free cyanide whilst RO is used to remove excess ammonia, nitrate and nitrite.

The clarified TSF effluent provided a good source of effluent to demonstrate the ability of the MBBR process to remove cyanide, ammonia, nitrite and nitrate in an industrial environment.

A pilot plant consisting of a 4 stage MBBR system was installed at the mine to demonstrate that biological treatment can be a viable treatment solution. Clarified effluent from the mine tailings pond containing residual free cyanide was used as influent wastewater to the 4 stage MBBR system designed for complete nitrogen removal. A process schematic is shown in Figure 7.

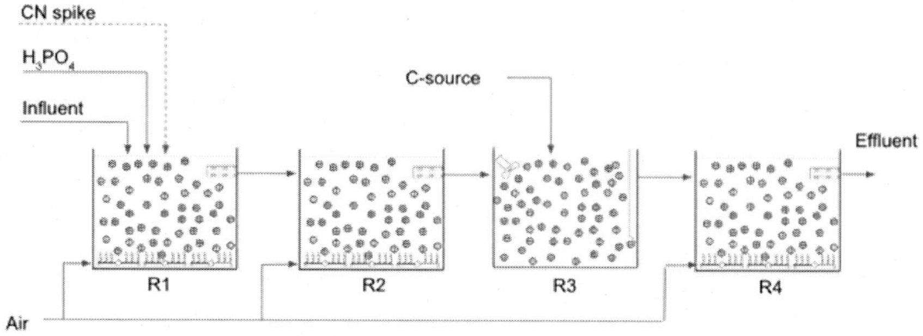

Figure 7. Process configuration for cyanide and nitrogen compound removal.

The process combines two aerobic MBBR, R1 and R2 with an anoxic MBBR, R3. Since cyanide is toxic to nitrifiers, it is advantageous to separate the cyanide oxidation step from the nitrification step in two reactors, R1 and R2. Post denitrification is undertaken in R3 to avoid having to recycle effluent from a nitrification step. The final aerobic stage, R4, is used for re-oxygenation before discharge.

Each reactor had an operating volume of 1 m^3. Clarified tailings pond effluent was fed to the MBBR at a controlled flowrate to maintain the hydraulic retention time within the MBBR between 1.3 and 3.5 hrs per stage. Phosphoric acid was added to provide a source of phosphorus for cellular synthesis. A source of organic carbon (ethanol or molasses) was added to the third stage MBBR designed for denitrification. Air was provided using a small compressor such that the dissolved oxygen concentration remained above 2 mg/L in the aerated reactors. The temperature of the effluent ranged from 26 to 30°C.

The performance of the MBBR was followed by daily spot measurements of free cyanide concentration in the effluent from R4. The discharge limit of 0.2 mg/L as free CN was met consistently; indeed, the average MBBR effluent free cyanide concentration was 0.016 mg/L for the test period between May and October 2021. The influent and effluent free cyanide concentrations to the MBBR are shown in Figure 8.

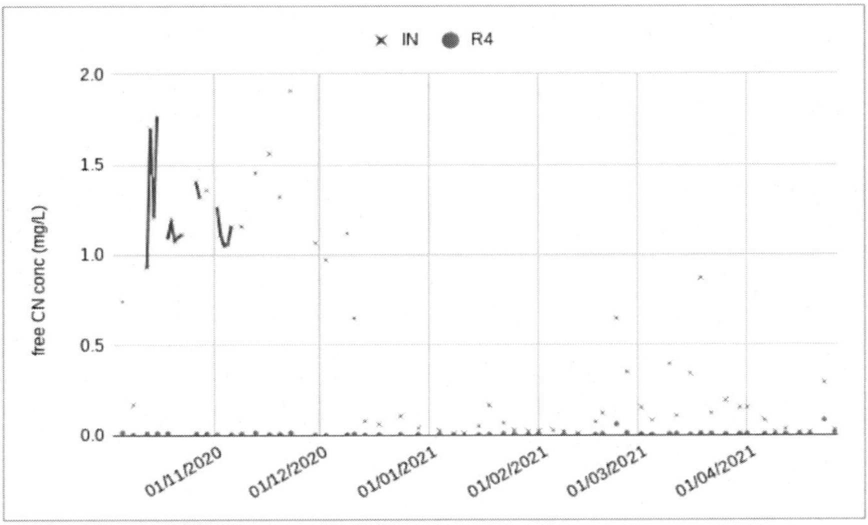

Figure 8. Clarified Tailing Pond (IN) and MBBR effluent (R4) free CN concentrations.

Since it is not infrequent for peak free cyanide concentrations to occur at mine sites, a test plan was put in place to demonstrate the robustness of the MBBR process. A concentrated sodium cyanide solution was prepared and dosed to the MBBR in addition to the clarified tailings pond effluent. The dose of NaCN was increased over time such that the influent concentration to the MBBR was increased in increments with the objective of maintaining the effluent concentration below 0.2 mg/L. The influent and effluent free CN concentrations during this test are shown in Figure 9.

Initially, residual CN was detected in the MBBR following the step increase - it can be seen from Figure 9 that a step increase from 2 mg/L to 3 mg/L free CN resulted in non-compliant effluent. Seeing no sign of improvement after more than 2 weeks at the new load, a carbon source (ethanol) was added to R1 for a few weeks to determine whether the kinetics of cyanide removal could be boosted by the addition of carbon as reported in literature (White et al., 2000). Without apparent improvement, the carbon dosing was stopped and the influent CN^- concentration was reduced to 2 mg/L. Once these changes were made, the final free CN concentration of 0.2 mg/L was achieved. The next incremental increases were limited to 0.5 mg/L to allow the MBBR to acclimatize to the increase in load until an influent CN concentration of 5 mg/L; the effluent from the MBBR remained below 0.2 mg/L throughout the remaining test period as can be seen on Figure 9.

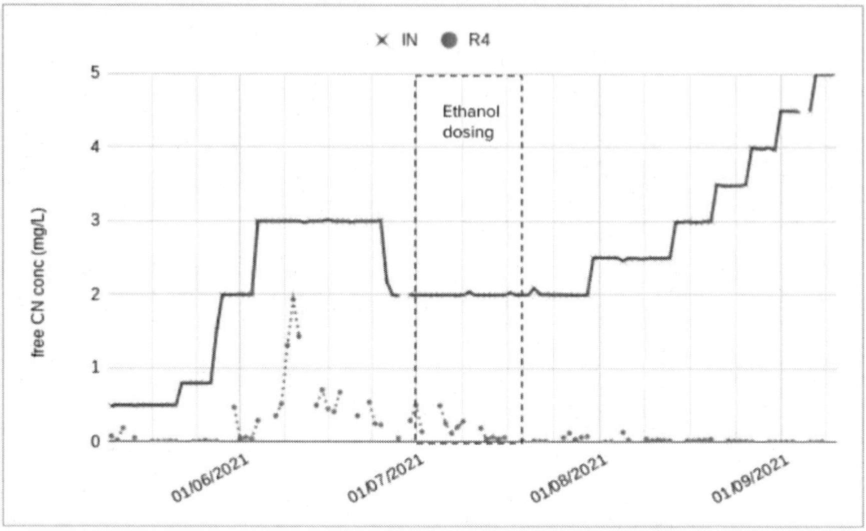

Figure 9. Clarified Tailing Pond effluent + NaCN spike (IN) and MBBR effluent (R4) free CN concentrations.

The results of the first phase of testing demonstrate that the MBBR can handle a gradual increase in load over time and produce an effluent compliant with the discharge limits for free cyanide - this data can be used in a startup phase to fully commission the biological system.

During normal operation of the mine, the cyanide concentration is likely to fluctuate unpredictably as a result of changing weather conditions or incidents in the process plant. A second phase of spiking events was initiated following the gradual incremental increases in load. In order to simulate a dysfunction or an incident upstream of a biological treatment plant, a series of 24hr spikes of greater amplitude was introduced. The spiking events were introduced following a period of low free CN concentration in the feed.

The 24 hrs spikes tested are presented in Table 3.

Table 3. Spiking program

Initial Conc (mg/L CN)	No of days at initial conc	Spike Conc (mg/L CN)
1	3	5
1	7	10
10	17	20
0.1	12	20

The results are presented in Figure 10. It can be seen that the first spikes were well tolerated by the MBBR and that the final effluent free CN concentration remained below 0.2 mg/L until the spike at 10 mg/L. Increasing the influent CN concentration to 10 mg/L resulted in a residual free CN concentration of 3 to 4 mg/L for 48 hrs. The influent free CN concentration was maintained at 10 mg/L until a compliant effluent was obtained and for an additional for 15 days before spiking at 20 mg/L. Interestingly, the final free CN concentration remained below 0.2 mg/L following the increase from 10 to 20 mg/L. The final free CN concentration remained below 0.2 mg/L at an influent concentration of 20 mg/L for a period of 12 days at a hydraulic retention time of 1.4hrs per reactor.

To simulate an incident or dysfunction in the process plant, the dosing of NaCN was stopped and the feed to the MBBR was solely the clarified tailings pond effluent for 12 days. A 24 hrs spike at 20 mg/L was then initiated. The free cyanide concentration was monitored every 4 hrs during the spiking event. During the spiking event, the free cyanide concentration increased to 14 mg/L at the outlet of the MBBR. The test was repeated 12 days later; again the cyanide removal efficiency dropped to 50% during the spiking event after an extended period of low load. The results of the spiking test are presented in Figure 10.

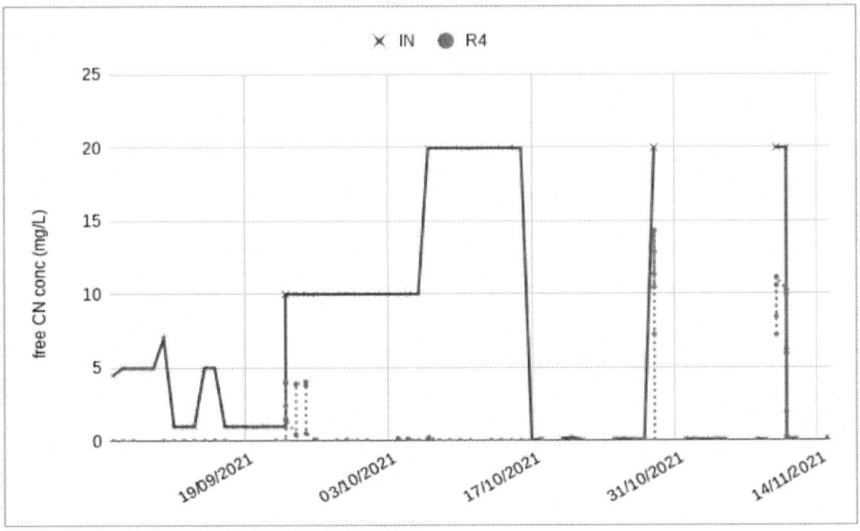

Figure 10. Clarified Tailing Pond effluent + NaCN spike (IN) and MBBR effluent (R4) free CN concentrations.

To determine the impact of increased free cyanide on the biological process, monitoring of nitrification and denitrification was also undertaken. Figure 11 shows the evolution of ammonia, nitrate and nitrite during the cyanide spiking test (continuous lines are influent concentrations whilst dashed lines are effluent concentrations). The data shows that gradual increases in cyanide are well tolerated however nitrification and denitrification are inhibited during the spiking event at 20 mg/L free cyanide. The system recovered within 24hrs after the end of the spiking event, with final ammonia, nitrate and nitrite concentrations returning to the initial levels thereby demonstrating the robustness of the MBBR process. With proper water management in the TSF, non-compliant effluent could be recycled to the TSF for a short period without impacting the level in TSF. Biological treatment is thus a viable option for TSF effluent treatment even in the event of incidental free cyanide discharge of up to 20 mg/L. Unfortunately, the tests could not be continued to evaluate the impact of larger spikes in free cyanide due to lack of resources. It would certainly have been interesting to determine the maximum allowable free cyanide concentration that would lead to complete failure of the biological process and the time required for recovery.

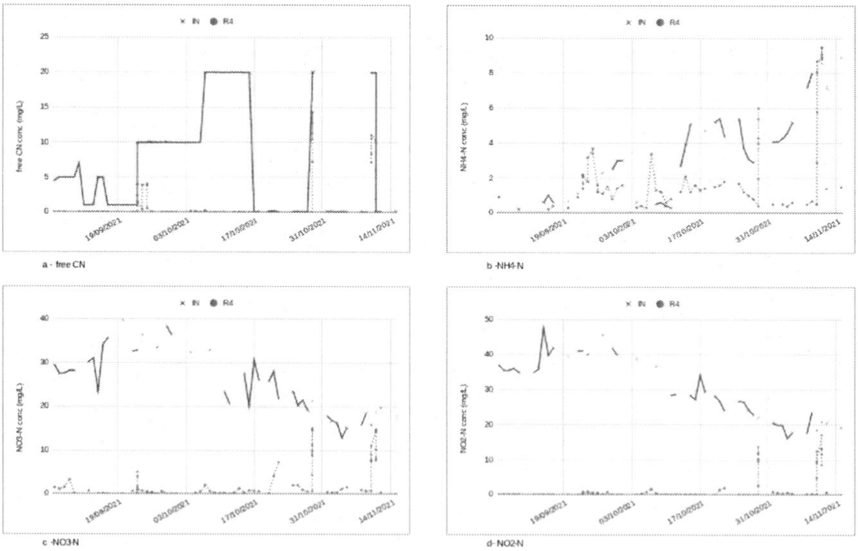

Figure 11. Evolution of free CN (a), NH4-N (b), NO3-N (c) and NO2-N (d) concentrations across MBBR plant during spiking events.

Conclusion

Cyanide is widely used within the gold mining industry and is not likely to be replaced in the near future. Treatment methods for cyanide removal are well established and allow the detoxification of mine effluents.

Innovations in biological treatment have widened the range of applications where biological treatment can be used - extensive testing at a mine has demonstrated that biological treatment is suitable for free cyanide removal and can handle variations in influent concentrations. Continuous operation for complete nitrogen removal at 20 mg/L free cyanide was demonstrated in a 4 stage MBBR system at hydraulic retention time of 1.4hrs per reactor. Spikes of 20 mg/L free cyanide resulted in inhibition of nitrification and denitrification for up to 24hrs. During this time, effluent can be recycled to the TSF without significant impact to the water balance of the site.

The capital costs and operating costs of these treatment systems should be evaluated on a case by case basis as the effluent characteristics, in particular free cyanide concentration, will have a significant impact on the selection of the most suitable treatment method and its operating cost.

References

Alvillo-Rivera A., Garrido-Hoyos S., Buitron G., Thangarasu-Sarasvati P. Rosano-Orteg G., Biological treatment for the degradation of cyanide: A review. *Journal of Materials Research and Technology* Vol 12, 2021 p 1418 -1433.

Aranguri, G., Reyes-Lopez, I., *Cyanide Degradation from Mining Effluent Using Two Reagents: Sodium Metabisulphite and the Metabisulphite Mixture with Hydrogen Peroxide, Tecciencia*, Vol. 13 No. 25, 1-9, 2018.

Botz, M. Overview of Cyanide Treatment Methods, Mining Environmental Management, *Mining Journal* Ltd., London, UK, 2001 May, pp. 28-30.

Botz M; Mudder T and Akcil A. Cyanide treatment: Physical, chemical and biological processes. *Advances in Gold Ore Processing* 2016, Chapter 36.

Ebbs Stephen. Biological degradation of cyanide compounds. *Biotechnology* 2004, 15:231–236.

EPA 600-3-76-038 Toxicity to fish of cyanides and other related compounds 1976.

EPA 530-R-94-037 Treatment of cyanide heap leaches and tailings 1984.

Habashi F. Kinetics and mechanism of gold and silver dissolution in cyanide solution. *Montana College of Mineral Science and Technology.* 1967 vol 59.

Hewitt D. Cyanide detoxification of cyanidation tails and process streams. *Mineral Processing and Extractive Metallurgy Volume* 121, 2012 p228-236.

Hilson, G., and Monhemius, A. J. Alternatives to cyanide in the gold mining industry: what prospects for the future? *Journal of Cleaner production,* 2006, Vol. 14, No. 12, pp. 1158-1167.

Ibrahim K., Syed M. Shukor M. and Ahmad S. Biological remediation of cyanide: a review. *Biotropia* Vol. 22 No. 2, 2015: 151 - 163.

ICMI International Cyanide Management Code for the Manufacture, Transport and Use of Cyanide in the Production of Gold www.cyanidecode.org.

Kim Y, Parka D, Leeb D., Park J. Inhibitory effects of toxic compounds on nitrification process for coke's wastewater treatment. *Journal of Hazardous Materials* 152 (2008) 915–921.

Kuyucaka N., Akcil A., Cyanide and removal options from effluents in gold mining and metallurgical processes. *Minerals Engineering* 50–51(2013)13–29.

Kwofie I., Jogand H., De Ladurantaye-Noël M. and Dale C. *Removal of Cyanide and Other Nitrogen-Based Compounds from Gold Mine Effluents Using Moving Bed Biofilm Reactor (MBBR) Water 2021*, 13, 3370.

Maier, R. M., Pepper, I. L., Gerba, J.P., 2009. *Environmental microbiology. In: Aquatic Environments: Case Study 6.1 – Beneficial Biofilm Removes Cyanide*, Elsevier Inc. Publication, pp. 111. (Chapter 6).

Mudder, T. I., M. M. Botz, and A. Smith. Chemistry and Treatment of Cyanidation Wastes, 2nd ed. London: *Mining Journal Books* Ltd., 2001.

Naveen D., Majumder C. B., Mondal P. and Dwivedi S. Biological Treatment of Cyanide Containing Wastewater Research. *Journal of Chemical Sciences* Vol. 1(7), 15-21, Oct. (2011).

Verbrugge B., Lanzano C. and Libassi M. The cyanide revolution: Efficiency gains and exclusion in artisanal- and small-scale gold mining. *Geoforum* Volume 126, November 2021, Pages 267-276.

Villemura R, Juteaub P., Bougiea V., Ménarda J., Déziel E. Development of four-stage moving bed biofilm reactor train with a pre-denitrification configuration for the removal of thiocyanate and cyanate. *Bioresource Technology* 181 (2015) 254–262.

White D., Pilon. T. and Woolard C. Biological treatment of cyanide containing wastewater. *Wat. Res.* Vol. 34, No. 7, pp. 2105±2109, 2000.

Chapter 5

Impact of Substrates on the Heat Capacity of Lyophilised Biomass of *Fusarium oxysporum* Associated with Cyanidation Wastewater

Enoch A. Akinpelu[1,2,*], PhD,
Seteno K. O. Ntwampe[3], PhD and Felix Nchu[1,2], PhD
[1]Bioresource Engineering Research Group (BioERG),
Cape Peninsula University of Technology, Cape Town, South Africa
[2]Department of Horticultural Sciences,
Cape Peninsula University of Technology, Bellville, South Africa
[3]Centre of Excellence in Carbon-Based Fuels,
School of Chemical and Minerals Engineering,
North-West University, Potchefstroom, South Africa

Abstract

Cyanide is a well-known constituent of mine wastewater that can be degraded by various processes. However, due to the cost and environmental challenges, microbial degradation seems to be the most effective process. When wastewater is treated with microorganisms, process performance should not only be based on toxicant degradation but also the impact of the toxicant on the physical properties of the microorganisms. The heat capacity of lyophilised biomass of *Fusarium oxysporum* was measured using modulated differential scanning calorimeter. The heat capacities for *F. oxysporum* grown in cyanidation wastewater were 1.1982, 1.077 and 1.143 J K^{-1} g^{-1} on glucose (GA), *Beta*

[*] Corresponding Author's E-mail: biyipelu@gmail.com.

In: Cyanide: Occurrence, Applications and Toxicity
Editor: Bill M. Torres
ISBN: 978-1-68507-619-1

vulgaris (BA) and cyanide supplemented with *Beta vulgaris* (BCN), respectively at 298.15 K and 1 atm. The enthalpies of formation of dry biomass are -297.58, -233.07 and -278.60 kJ/C-mol for BA, BCN and GA, respectively. These values were found to be within the range of some biological molecules. The presence of cyanide in the wastewater minimally affected the thermodynamic property of the dried biomass of *F. oxysporum*.

Keywords: *Beta vulgaris*, biodegradation, cyanide, heat capacity, *Fusarium oxysporum*

Introduction

There are several reports on the microbial treatment of cyanidation wastewater with successful operational efficiency (Akinpelu et al., 2016b, Mekuto et al., 2017, Virender Kumar & Bhalla, 2013, Ibrahim et al., 2016). The impact of various factors such as temperature, pH, substrate and bioreactor configuration on the performance of organisms in cyanidation wastewater have been reported (Campos et al., 2006, Luque-Almagro et al., 2016, Mekuto et al., 2015). In addition, thermodynamic tools have been used to validate microbial performance and cyanide degradation (Akinpelu et al., 2016a, Behnamfard & Salarirad, 2009, Singh & Balomajumder, 2016). However, the uptake of the microbial remediation by the industry have not been inspiring even though Homestake and LaRonde gold mine in Canada, including Gold Fields Limited have demonstrated the robustness and feasibility of the biological process (Huddy et al., 2015, Stott et al., 2001, Du Plessis et al., 2001). Analysis of process performance should not only be based on microbial growth and toxicant degradation but also the impact of the toxicant on the physical properties of the organism. This will provide further insight into the capability of the organism to manage its environment under stress.

Thermodynamic properties of a material are an essential tool for predicting the feasibility of any chemical and biological reaction including processes such as the microbial growth process and the biomass conversion of nutrient media to useful products. Among these thermodynamic properties, the heat capacity of biological molecules such as starch, glucose, proteins and amino acids reportedly measured based on rudimentary heat capacity quantifications can be used to estimate entropy increments and/or changes at low temperatures (0 to 298.15 K); however, there is high uncertainty associated with this estimate (Boerio-Goates, 1991). Recently, some researchers have reported on the use of an adiabatic calorimeter for measuring heat capacity of biological materials at low temperature; based on the application of the third law of thermodynamics, for which incremental entropy and/or absolute entropy can be estimated (Pyda, 2001, Kabo et al., 2013). Nevertheless, the results were determined to be unreproducible because there was no reference material used and that each researcher had to fabricate their calorimeter. This may be the reason for Pyda's (Pyda, 2001) preference for Differential Scanning Calorimeter (DSC) measurements over adiabatic calorimeter measurements. Overall, there is only one report on the heat capacity of microbial dried biomass thus far (Battley et al., 1997) which reported on the use of an adiabatic calorimeter for quantifying the heat capacity of lyophilised cells of *Saccharomyces cerevisiae,* after the estimation of entropy changes as a function of temperature based on the third law of thermodynamics.

From the second law of thermodynamics, the heat capacity of any material can be estimated/quantified using heat flow curves obtained from DSC generated profiles of the sample being studied. A DSC provides more reliable, accurate and reproducible results because it is calibrated with a standard reference and/or material such as sapphire, which is used to ascertain and/or detect any error with the equipment, a parametric requirement not available with an adiabatic calorimeter. Nevertheless, it is often difficult to interpret the heat flow data from DSC experiments when multiple processes are involved, resulting in overlapping transitions. Besides, the heat capacity of a material cannot be determined directly from DSC data, it requires multiple experiments including data interpretation to ascertain or determine the heat capacity (Verdonck et al., 1999, Brantley et al., 2003, Xie et al., 2010, Magoń & Pyda, 2013).

Furthermore, a modulated DSC (MDSC™) overcomes these drawbacks and thus provide an insight into the thermal properties of materials being studied. MDSC™ uses a modulated temperature input signal to provide information on the heat capacity, both under isothermal and non-isothermal conditions. Further details on theory, principles, application and instrumentation requirements of the MDSC™ can be found in Vendonck (Verdonck et al., 1999) and Knopp (Knopp et al., 2016).

Cells are known to be insoluble and using lyophilised cells, an approximate entropy per unit mass can be obtained if cellular integrity is not annihilated by lyophilisation (Battley, 1987).

Battley (Battley, 1999) in his report hypothesized that the constituent materials of formation do not affect the molecular weight and subsequently thermodynamic properties of lyophilised cells but Duboc et al. (Duboc et al., 1999) and Akinpelu et al. (Akinpelu et al., 2018) have proven otherwise with their report on yeast, bacteria, algae and filamentous fungi. And this agrees with the basic concept that the specific heat is a function of the property of the substance. Therefore, the objective of this study was to determine the impact of different substrates (glucose, *Beta vulgaris* and cyanide) on the heat capacity of lyophilised biomass of *Fusarium oxysporum* in cyanidation wastewater using a modulated DSC (MDSC™).

Method

Sample Preparation

Carbon sources; glucose (GA), *Beta vulgaris* (BA), and *Beta vulgaris* with cyanide (BCN) were used for the cultivation of *Fusarium oxysporum* in gold mine wastewater as shown in (Akinpelu et al., 2018). Briefly, *Fusarium oxysporum* was grown in a 1 litre continuously stirred tank reactor (CSTR) containing metal-laden synthetic gold mine wastewater at 25 ± 2°C. The wastewater contains (per litre): 47 mg $CuSO_4\ 5H_2O$, 42 mg $Fe_2(SO_4)_3\ H_2O$, 278 mg $(NH_4)_2SO_4$, 27mg KH_2PO_4, 3mg $ZnSO_4\ 7H_2O$, 0.9 mg $PbBr_2$, and 40 mg $Na_2HAsO_4\ 7H_2O$. The reactor was inoculated with 10% (v/v) *F. oxysporum* and 0.3 g glucose as refined carbon sources. Subsequently, experiments on 0.3 g pulverised *B. vulgaris* followed by 0.3 g pulverised *B. vulgaris* with 100 ppm CN^-/L in form of KCN. Biomass was harvested once the carbon source was used up and/or when the microbial growth reached the stationary phase. The harvested biomass was centrifuge at 4°C for 10 min at a speed of 10,000rpm in an Avanti® J-E centrifuge (Beckman Coulter, Inc. USA). Recovered biomass was washed thrice in sterile distilled water, dried in a Duran® vacuum desiccator (DURAN Group GmbH, Germany), and stored at -20°C for further analyses.

Sample dried biomass was dissolved in sterile distilled water in a 1:1, weight: volume ratio and incubated at 298.15 K for 16 h to ferment any residual carbohydrates. The Durham tube method was used to test for any residual carbohydrates in the suspension (Battley, 1999).

The procedures were repeated until an appropriate quantity of dry biomass was obtained. The elemental analysis of the samples was determined by a Thermo Flash EA 1112 series analyser in a Helium carrier gas (Thermo Fisher Scientific Inc. Waltham, USA) after estimation of the molar mass of dried biomass containing a unit carbon as 23.03, 33.14 and 27.06 g/C-mol for GA, BA and BCN samples, respectively. The detailed elemental analysis of the samples is presented in (Akinpelu et al., 2018).

Combustion Calorimetry

The heat of combustion of the samples was determined in an e2k Bomb calorimeter (Digital Data Systems Pty Ltd, South Africa) in triplicates as described in (Akinpelu et al., 2018). Briefly, a pre-cut firing cotton (Part No. CAL2K-4-FC) was looped over the firing wire (Part No. CAL2K-4-FW) and twisted at the ends. A crucible containing 0.30 g of dried biomass was inserted into the outside electrode's crucible holder, with the firing cotton touching the sample. The electrode assembly was loaded into the vessel body and filled with 3000 kPa oxygen. The vessel was removed from the filling station and allowed to stabilize for 1 min before insertion into the calorimeter. The analytical grade Benzoic acid (Part No. CAL2K-BA) was used for calorimeter calibration.

Modulated Differential Scanning Calorimeter (MDSC)

The MDSC™ uses both the linear heating rate as well as a modulated (sinusoidal) heating rate to determine the total heat flow rate, see Eq. 1. The linear heating rate provides information on the total heat rate while the sinusoidal heating rate provides the heat capacity information from a fraction of the heat flow (Stark et al., 2013).

$$\frac{dQ}{dt} = C_P\beta + f(T,t) \qquad (1)$$

where (dQ/dt) is the total heat flow due to the linear heating rate (the equivalent of standard DSC), C_P is the heat capacity component calculated from the heat flow due to the sinusoidal heating rate, β is the linear heating rate of the sample, $C_P\beta$ is the reversing heat flow while $f(T,t)$ is the kinetic component of total heat flow known as the non-reversing heat flow which can be calculated from the difference between the reversing heat flow and total heat signal component.

In this study, a Discovery DSC® (TA Instruments, Inc. New Castle, DE, USA) equipped with a modulated Differential Scanning Calorimeter (MDSC™) software using a Liquid Nitrogen Cooling Accessory (LNCA) at atmospheric pressure was used for all measurements. Helium gas at a flow rate of 50 mL/min was purged through the MDSC™. The MDSC™ was equilibrated at a temperature of -150°C and measurements were performed at an underlying heating rate of 3 K/min up to a temperature of 100°C; the amplitude of modulation 1 K and period of 60 s for the sample mass of 2 mg – dry biomass weight. Temperature and heat capacity calibrations were performed with an MDSC™ certified Indium reference material (Part No. 915061.901) and a sapphire for the specific heat capacity determinations (Part No. 9703790.901), respectively at similar operating conditions as used for the samples. The data were analysed using a TRIOS software v4.1.1.33073 (TA Instruments Inc. USA). All procedures were done in triplicate. All datasets are openly available in (Akinpelu et al., 2017).

Results and Discussion

Fusarium oxysporum Growth on Substrates

Figure 1 showed *Fusarium oxysporum* experienced a lag phase on all substrates which lasted for almost 48 hours on sample containing cyanide (BCN) and can be attributed to the inhibitory effect of cyanide on the medium. The growth on beetroot (BA) was conspicuously higher than that on refined carbon (GA) and the sample containing cyanide (BCN). The availability of numerous soluble sugars in beetroot including the beetroot processing method aided the microbial growth in sample BA (Wruss et al., 2015). In addition, the *F. oxysporum.* growth on substrate BA, showed a diauxic pattern that utilises more than one substrate for its metabolism.

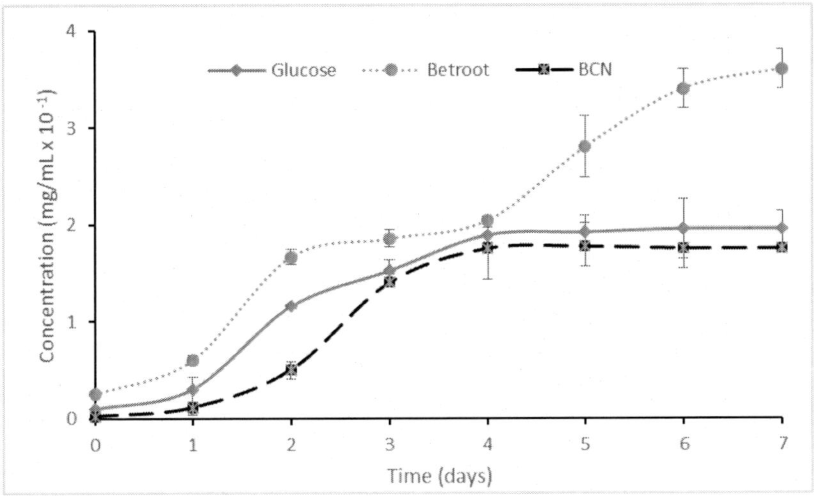

Figure 1. *Fusarium oxysporum* growth on different substrates.

Phase Transition

MDSC^{TM} allows for separation and evaluation of thermodynamic and kinetic processes within the regions of glass and melting transitions as shown in equation (1). Endothermic or exothermic enthalpy relaxation may occur within the glass transition range owing to changes in the temperature of the sample. There was no glass transition (T_g) on total heat flow and non-reversible heat flow profiles except on reversible heat flow profiles for all biomass samples tested. There was glass transition (T_g) at a temperature of 239 K and the enthalpy change of 12.586 J/g with an endothermic peak temperature of 292.333 K for BA sample, and a T_g of 211 K, an enthalpy change of 22.096 J/g at an endothermic peak temperature of 287 K for BCN samples on the reversible heat flow profiles which was an indication of a structural transformation of the samples studied, i.e., BA and BCN samples; that is reversible with any temperature changes (Lai & Lii, 1999, Tan et al., 2004). The lower T_g for BCN samples is an indication of a rapid breakdown of aromatic constituents in *Beta vulgaris* supplemented biomass owning to the residual free cyanide within the synthetic wastewater used, thereby enhancing microbial growth during the free cyanide biodegradation process. In addition, melting transition (T_m) was observed on the reversing heat flow profiles for all samples studied – see Table 1.

Table 1. Effect of melting temperature on the samples

Samples	T_m (onset) (K)	T_m (peak) (K)	ΔH (J/g)
GA	218.72 ± 1.05	277.11 ± 0.90	17.48 ± 1.01
BA	212.31 ± 1.02	292.33 ± 1.01	12.59 ± 0.80
BCN	127.34 ± 1.04	287.73 ± 1.03	22.09 ± 1.05

Microbial degradation is directly affected by T_m, i.e., the lower the T_m, the higher the biodegradation of a toxicant such as free cyanide (Herzog et al., 2006, Mueller, 2006, Kasuya et al., 2009). *Beta vulgaris* consist of 9.56% carbohydrates, with betalains, phenolic compounds, including trace elements and minerals accounting for a larger percentage (Wruss et al., 2015, USDA, 2016). The presence of betalains and phenolic compounds influence the microbial growth in BA samples since they degrade under different bioreactor operating conditions (Herbach et al., 2006) from that of free cyanide, resulting in a slightly higher molecular weight of the *F. oxysporum* and thus, a higher T_m for BA samples. Similarly, interactions of molecular chains affect the change of enthalpy (ΔH) in melting with the internal energy accounting for the flexibility or otherwise of the samples studied, thus affecting the change of entropy (ΔS) in melting (Tokiwa et al., 2009). The highest ΔH in melting for BCN samples was an indication of the highest molecular interactions during free cyanide biodegradation.

Tan et al. (Tan et al., 2004) reported glass transition on reversible heat flow profile of MDSC™ after the onset temperature which overlapped with the peak temperature. It was presumed that this was due to changes in the state of the starch molecules i.e., from being highly confined within the granular packing, to being disentangled as the order in which the molecules are arranged changed as the transition occurred. Glass transition could also be due to structural transitions of cellular materials at temperatures below the freezing point of pure water including the influence of the underlying heating rate, modulation period and amplitude (Battley et al., 1997, Stark et al., 2013). There was no glass transition on *F. oxysporum* grown on glucose i.e., GA, this is similar to the previous report on the heat capacity of starch and glucose (Kabo et al., 2013, Boerio-Goates, 1991). It was evident that the substrate from which the biomass materials were formed, played a major role in the phase transition as can be seen that only cultures grown on *Beta vulgaris* (BA and BCN), had a glass transition.

Heat Capacity

Heat capacity measurements were obtained from 130 to 305 K. The experimental result of the specific heat capacity of Sapphire indicated a general deviation within ± 2 percent of standard value as shown in Table 2.

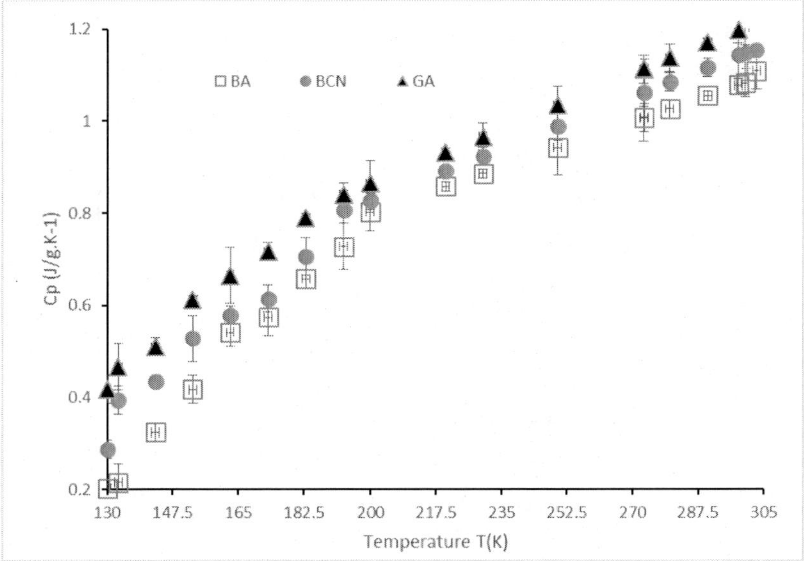

Figure 2. Heat capacity of samples as a function of temperature.

An indication that the experimental procedure and results are reliable. The experimental specific heat capacity of lyophilised cells of *F. oxysporum* samples in the temperature range of 130 to 305 K are presented in Figure 2.

Although there are no general statements on heat capacity changes, the heat capacity of the samples increased steadily with a temperature rise. The heat capacity shows a linear relationship with temperature over the range of (190 to 300) K as shown in equations (2), (3) and (4), respectively.

$$BA\colon C_p(J.K^{-1}g^{-1}) = 0.2422 + 0.0028\,T\,(K) \tag{2}$$

$$BCN\colon C_p(J.K^{-1}g^{-1}) = 0.1889 + 0.0032\,T\,(K) \tag{3}$$

$$GA\colon C_p(J.K^{-1}g^{-1}) = 0.1845 + 0.0034\,T\,(K) \tag{4}$$

Table 2. Comparison of standard and experimental heat capacity of Sapphire

Temperature (K)	**Literature value ($JK^{-1}g^{-1}$) (Ditmars et al., 1982)**	**Experimental value ($JK^{-1}g^{-1}$)**	**Deviation (%)**
130	0.2349	0.2305 ± 0.02	-1.873
140	0.2739	0.2685 ± 0.01	-1.971
150	0.3134	0.3142 ± 0.04	+0.255
160	0.3526	0.3530 ± 0.05	+0.113
170	0.3913	0.3909 ± 0.02	-0.102
180	0.4291	0.4218 ± 0.03	-1.701
190	0.4659	0.4662 ± 0.01	+0.064
200	0.5014	0.4999 ± 0.01	-0.299
220	0.5684	0.5679 ± 0.05	-0.088
230	0.5996	0.6001 ± 0.04	+0.083
250	0.6579	0.6505 ± 0.04	-1.125
270	0.7103	0.7120 ± 0.02	+0.239
273.15	0.7180	0.7167 ± 0.03	-0.181
280	0.7343	0.7353 ± 0.02	+0.136
290	0.7572	0.7561 ± 0.01	-0.145
300	0.7788	0.7775 ± 0.03	-0.167

Table 3. Comparison of specific heat capacities ($J\ K^{-1}g^{-1}$) for organic samples

Temp (K)	**BA (This study)**	**BCN (This study)**	**GA (This study)**	**Starch[a]**	**Yeast[b]**	**Glucose[c]**	**Starch[d]**
200	0.80 ± 0.04	0.83 ± 0.02	0.86 ± 0.05	0.73	0.90	0.82	0.76
250	0.94 ± 0.06	0.99 ± 0.03	1.03 ± 0.04	0.93	1.11	1.02	0.95
298.15	1.07 ± 0.02	1.14 ± 0.05	1.19 ± 0.03	1.14	1.30	1.21	1.15
300	1.08 ± 0.03	1.15 ± 0.06	1.20 ± 0.04	1.15	1.31	1.22	1.16

a, b, c & d represent Kabo et al., 2013, Battley et al., 1997, Boerio-Goates, 1991 & Pyda, 2001, respectively.

The heat capacity of GA samples was shown to be the highest. The low heat capacity observed for BCN samples may be due to the stress imposed by the free cyanide during microbial proliferation which in turn could affect the biomass structural integrity compared to GA samples. Furthermore, BA samples recorded the lowest heat capacity when compared with all other samples studied. This was expected since *Beta vulgaris* contains approximately 9.56% carbohydrate that served as an energy and/or carbon source for the *F. oxysporum* used (Wruss et al., 2015, USDA, 2016). The higher values of heat capacity in BCN samples when compared to BA samples could be directly related to the microorganism's ability to utilise cyanide as a carbon source to supplement carbohydrates in *B. vulgaris* agro-industrial waste. Since there is no other reports on heat capacities of microbial dried biomass in literature particularly for fungi, comparing the heat capacity of *Fusarium oxysporum* with *Saccharomyces cerevisiae* showed that from the results obtained, GA samples were comparatively similar to those reported by Battley (Battley et al., 1997).

Table 3 compares the values of specific heat capacities of starch and/or glucose samples reported by previous authors. At a temperature range of (200 to 300) K, our result is closer to the Boerio-Goates report (Boerio-Goates, 1991) especially GA sample. All the specific heat capacities reported by (Battley et al., 1997) were higher than our findings while (Kabo et al., 2013) is lower than our reports for BCN and GA samples. Generally, our report can be seen within the reported heat capacity for starch in the literature.

Although, a previous study on the propagation of *Fusarium oxysporum* showed a preference for *Beta vulgaris* with substantial free cyanide biodegradation including diauxic growth (Akinpelu et al., 2016b). Battley (Battley et al., 1997) acknowledged that the results are not reproducible, due to the use of an adiabatic calorimeter. However, the results reported herein; i.e., from an MDSC™ are generally deemed reliable thus reproducible. Albeit, if sufficient quantities of *B. vulgaris* and/or further pre-treatment of agro-industrial waste besides pulverising were used, better biodegradation and substrate utilisation performance could have been observed, as this influenced the integrity and the quality of the biomass being generated thus bioreactor performance. As such, it is feasible to suggest that a highly stressful environment culminates in poorly structured cells, which can impede the ability of such cells to detoxify a highly contaminated environment.

Elemental and Combustion Analysis

The elemental analysis of the dried biomass samples in percentages of *C*, *H*, *N*, and *O* were determined to give an empirical formula representing the energy constituent of the biomass. Phosphorus and sulphur are not normally included because they do not play an important role in the material balance of cellular reactions and their oxides are part of the ash formed on combustion, thus they are ignored in the empirical formulation (Battley et al., 1997).

Table 4. Enthalpy of combustion and formation of dry biomass at 298.15 K and 1 atm

Microorganisms	**Elemental formula**	$\Delta H_C^o(KJ/C - mol)$	$\Delta H_f^o(KJ/C - mol)$	**Reference**
BA	$CH_{2.377}N_{0.091}O_{1.093}$	-435.78 ± 1.04	-297.58 ± 1.01	This study
BCN	$CH_{1.82}N_{0.027}O_{0.804}$	-420.50 ± 1.76	-233.07 ± 0.98	This study
GA	$CH_{1.167}N_{0.067}O_{0.558}$	-281.69 ± 0.47	-278.60 ± 1.02	This study
A. niger	$CH_{1.5}N_{0.12}O_{0.53}$	-418.7		(Duboc et al., 1999)
Rocan I	$CH_{1.4}N_{0.04}O_{0.5}$	-473.3		(Duboc et al., 1999)
B. flavum	$CH_{1.8}N_{0.19}O_{0.33}$	-491.7		(Duboc et al., 1999)
S. cerevisiae	$CH_{1.613}N_{0.158}O_{0.557}$	-509.37	-133.13	(Battley, 1999)

The empirical formula of the biomass containing a unit carbon for BA, BCN and GA samples are $CH_{2.377}N_{0.091}O_{1.093}$, $CH_{1.82}N_{0.027}O_{0.804}$, and $CH_{1.167}N_{0.067}O_{0.558}$, with molecular weight of 33.14, 27.06 and 23.03 g/C-mol, respectively (Akinpelu et al., 2018) which aligns with the previous report for dry biomass (Battley, 1999, Duboc et al., 1999). The enthalpy of combustion (ΔH_C^o) of dry biomass in bomb calorimeter is presented in Table 4 based on the reactions (5), (6) and (7) below:

$$BA: CH_{2.377}N_{0.091}O_{1.093}(cell) + 1.048O_{2(g)} \rightarrow CO_{2(g)} + 1.189H_2O_{(l)} + 0.046N_{2(g)} \quad (5)$$

$$BCN: CH_{1.82}N_{0.027}O_{0.804}(cell) + 1.053O_{2(g)} \rightarrow CO_{2(g)} + 0.91H_2O_{(l)} + 0.0135N_{2(g)} \quad (6)$$

$$GA: CH_{1.167}N_{0.067}O_{0.558}(cell) + 1.0127O_{2(g)} \rightarrow CO_{2(g)} + 0.5835H_2O_{(l)} + 0.0335N_{2(g)} \quad (7)$$

The enthalpy of combustion of the samples agrees with the previous reports except for the sample GA. The reason for this deviation is not clear. The enthalpy of formation (ΔH_f^o) for each sample was calculated using standard heat of formation $\Delta H_f^o(CO_{2(g)}) = -393.51\ kJ/mol$ and $\Delta H_f^o(H_2O_{(l)}) = -285.83\ kJ/mol$ (Battley, 1999), according to the reactions (5), (6) and (7).

The accuracy of these values (ΔH_f^o) is a function of the validity of the molecular formula of the carbohydrate used to represent the *B. vulgaris* agro-waste which has a direct influence on the accuracy of the heat determination. Environmental factors such as temperature, pH, substrate concentration and presence of toxins have been used extensively to elucidate the feasibility of microbial degradation. Estimation of physical properties such as heat capacity and heat of formation will further ascertain the veracity of microorganisms in environmental engineering applications.

Conclusion

The results presented herein showed that the substrate used for microbial growth can affect the determination of the heat capacity as well as enthalpy of biological samples. The determination of such heat capacity of a material as a function of the structure of the material can assist in substrate selection for cyanidation wastewater treatment. The impairment caused by free cyanide is also reflected in the heat capacity and enthalpy of dry biomass assessed. The toxicity by free cyanide reduces the biomass molecular degree of freedom hence, the biomass could not store sufficient thermal energy as most energy resources are dedicated to cellular maintenance. The enthalpy of combustion of dried biomass of *F. oxysporum* is within the range available in the literature. This suggests that the method of biomass preparation and its constituents do not significantly affect the final enthalpy of biomass formation. Since biomass is not a completely crystalline substance, the enthalpy derived from the calorimetric measurement can be used to further elucidate the capabilities associated with the novel biocatalyst selection for the bioremediation of cyanidation wastewater.

Acknowledgments

The authors acknowledge Ross Burnham of Advanced Laboratory Solutions for his assistance on DSC data interpretation and the funding from the Cape Peninsula University of Technology (CPUT), University Research Fund (URF RK 16).

References

Akinpelu, E., Ntwampe, S., Mekuto, L. & Ojumu, T. 2017. Thermodynamic Data of *Fusarium oxysporum* Grown on Different Substrates in Gold Mine Wastewater. *Data,* 2, 24.

Akinpelu, E. A., Ntwampe, S. K., Mpongwana, N., Nchu, F. & Ojumu, T. V. 2016a. Biodegradation Kinetics of Free Cyanide in *Fusarium oxysporum-Beta vulgaris* Waste-metal (As, Cu, Fe, Pb, Zn) Cultures under Alkaline Conditions. *BioResources,* 11, 2470-2482.

Akinpelu, E. A., Ntwampe, S. K. O. & Chen, B.-H. 2018. Biological stoichiometry and bioenergetics of *Fusarium oxysporum* EKT01/02 proliferation using different substrates in cyanidation wastewater. *The Canadian Journal of Chemical Engineering,* 96, 537-544.

Akinpelu, E. A., Ntwampe, S. K. O., Mekuto, L. & Itoba Tombo, E. F. 2016b. Optimizing the bioremediation of free cyanide containing wastewater by *Fusarium oxysporum* grown on beetroot waste using response surface methodology. In: Ao, S. I., Douglas, C. & Grundfest, W. S. (eds.) *Lecture Notes in Engineering and Computer Science: Proceedings of the World Congress on Engineering and Computer Science.* San Francisco, USA: Newswood Limited, 664-670.

Battley, E. H. 1987. *Energetics of microbial growth,* New York, Wiley Inter-science.

Battley, E. H. 1999. The thermodynamics of microbial growth. *In:* Kemp, R. B. (ed.) *Handbook of thermal analysis and calorimetry.* Amsterdam: Elsevier, 219-266.

Battley, E. H., Putnam, R. L. & Boerio-Goates, J. 1997. Heat capacity measurements from 10 to 300 K and derived thermodynamic functions of lyophilized cells of Saccharomyces cerevisiae including the absolute entropy and the entropy of formation at 298.15 K. *Thermochimica Acta,* 298, 37-46.

Behnamfard, A. & Salarirad, M. M. 2009. Equilibrium and kinetic studies on free cyanide adsorption from aqueous solution by activated carbon. *Journal of Hazardous Materials,* 170, 127-133.

Boerio-Goates, J. 1991. Heat-capacity measurements and thermodynamic functions of crystalline α-D-glucose at temperatures from 10 K to 340 K. *The Journal of Chemical Thermodynamics,* 23, 403-409.

Brantley, W. A., Iijima, M. & Grentzer, T. H. 2003. Temperature-modulated DSC provides new insight about nickel-titanium wire transformations. *American Journal of Orthodontics and Dentofacial Orthopedics,* 124, 387-394.

Campos, M. G., Pereira, P. & Roseiro, J. C. 2006. Packed-bed reactor for the integrated biodegradation of cyanide and formamide by immobilised *Fusarium oxysporum* CCMI 876 and *Methylobacterium* sp. RXM CCMI 908. *Enzyme and Microbial Technology,* 38, 848-854.

Ditmars, D., Ishihara, S., Chang, S., Bernstein, G. & West, E. 1982. Enthalpy and heat-capacity standard reference material: synthetic sapphire (α-Al_2O_3) from 10 to 2250 K. *J. Res. Natl. Bur. Stand,* 87, 159-163.

Du Plessis, C., Barnard, P., Muhlbauer, R. & Naldrett, K. 2001. Empirical model for the autotrophic biodegradation of thiocyanate in an activated sludge reactor. *Letters in Applied Microbiology,* 32, 103-107.

Duboc, P., Marison, I. & Von Stockar, U. 1999. Quantitative calorimetry and biochemical engineering. In: Kemp, R. B. (ed.) *Handbook of thermal analysis and calorimetry.* Amsterdam: Elsevier, 267-365.

Herbach, K. M., Stintzing, F. C. & Carle, R. 2006. Betalain Stability and Degradation—Structural and Chromatic Aspects. *Journal of Food Science,* 71, R41-R50.

Herzog, K., Müller, R. J. & Deckwer, W. D. 2006. Mechanism and kinetics of the enzymatic hydrolysis of polyester nanoparticles by lipases. *Polymer Degradation and Stability,* 91, 2486-2498.

Huddy, R. J., van Zyl, A. W., van Hille, R. P. & Harrison, S. T. L. 2015. Characterisation of the complex microbial community associated with the ASTER™ thiocyanate biodegradation system. *Minerals Engineering,* 76, 65-71.

Ibrahim, K. K., Syed, M. A., Shukor, M. Y. & Ahmad, S. A. 2016. Biological Remediation of Cyanide: A Review. *BIOTROPIA-The Southeast Asian Journal of Tropical Biology,* 22, 151-163.

Kabo, G. J., Voitkevich, O. V., Blokhin, A. V., Kohut, S. V., Stepurko, E. N. & Paulechka, Y. U. 2013. Thermodynamic properties of starch and glucose. *The Journal of Chemical Thermodynamics,* 59, 87-93.

Kasuya, K. I., Ishii, N., Inoue, Y., Yazawa, K., Tagaya, T., Yotsumoto, T., Kazahaya, J. I. & Nagai, D. 2009. Characterization of a mesophilic aliphatic–aromatic copolyester-degrading fungus. *Polymer Degradation and Stability,* 94, 1190-1196.

Knopp, M. M., Löbmann, K., Elder, D. P., Rades, T. & Holm, R. 2016. Recent advances and potential applications of modulated differential scanning calorimetry (mDSC) in drug development. *European Journal of Pharmaceutical Sciences,* 87, 164-173.

Lai, V. F. & Lii, C. Y. 1999. Effects of modulated differential scanning calorimetry (MDSC) variables on thermodynamic and kinetic characteristics during gelatinization of waxy rice starch. *Cereal chemistry,* 76, 519-525.

Luque-Almagro, V. M., Moreno-Vivián, C. & Roldán, M. D. 2016. Biodegradation of cyanide wastes from mining and jewellery industries. *Current Opinion in Biotechnology,* 38, 9-13.

Magoń, A. & Pyda, M. 2013. Apparent heat capacity measurements and thermodynamic functions of d(−)-fructose by standard and temperature-modulated calorimetry. *The Journal of Chemical Thermodynamics,* 56, 67-82.

Mekuto, L., Ntwampe, S. & Jackson, V. 2015. Biodegradation of free cyanide and subsequent utilisation of biodegradation by-products by Bacillus consortia: optimisation using response surface methodology. *Environmental Science and Pollution Research,* 22, 10434-10443.

Mekuto, L., Ntwampe, S. K. O., Utomi, C. E., Mobo, M., Mudumbi, J. B., Ngongang, M. M. & Akinpelu, E. A. 2017. Performance of a continuously stirred tank bioreactor system connected in series for the biodegradation of thiocyanate and free cyanide. *Journal of Environmental Chemical Engineering,* 5, 1936-1945.

Mueller, R. J. 2006. Biological degradation of synthetic polyesters—Enzymes as potential catalysts for polyester recycling. *Process Biochemistry,* 41, 2124-2128.

Pyda, M. 2001. Conformational contribution to the heat capacity of the starch and water system. *Journal of Polymer Science Part B: Polymer Physics,* 39, 3038-3054.

Singh, N. & Balomajumder, C. 2016. Equilibrium isotherm and kinetic studies for the simultaneous removal of phenol and cyanide by use of *S. odorifera* (MTCC 5700) immobilized on coconut shell activated carbon. *Applied Water Science*, 1-15.

Stark, W., Jaunich, M. & McHugh, J. 2013. Cure state detection for pre-cured carbon-fibre epoxy prepreg (CFC) using Temperature-Modulated Differential Scanning Calorimetry (TMDSC). *Polymer Testing,* 32, 1261-1272.

Stott, M. B., Franzmann, P. D., Zappia, L. R., Watling, H. R., Quan, L. P., Clark, B. J., Houchin, M. R., Miller, P. C. & Williams, T. L. 2001. Thiocyanate removal from saline CIP process water by a rotating biological contactor, with reuse of the water for bioleaching. *Hydrometallurgy,* 62, 93-105.

Tan, I., Wee, C. C., Sopade, P. A. & Halley, P. J. 2004. Investigation of the starch gelatinisation phenomena in water–glycerol systems: application of modulated temperature differential scanning calorimetry. *Carbohydrate Polymers,* 58, 191-204.

Tokiwa, Y., Calabia, B., Ugwu, C. & Aiba, S. 2009. Biodegradability of Plastics. *International Journal of Molecular Sciences,* 10, 3722.

USDA 2016. USDA National Nutrient Database for Standard Reference Release 28. September, 2015 ed. USA: USDA.

Verdonck, E., Schaap, K. & Thomas, L. C. 1999. A discussion of the principles and applications of Modulated Temperature DSC (MTDSC). *International Journal of Pharmaceutics,* 192, 3-20.

Virender Kumar, V. K. & Bhalla, T. C. 2013. In vitro cyanide degradation by Serretia marcescens RL2b. *Int J Environ Sci,* 3.

Wruss, J., Waldenberger, G., Huemer, S., Uygun, P., Lanzerstorfer, P., Müller, U., Höglinger, O. & Weghuber, J. 2015. Compositional characteristics of commercial beetroot products and beetroot juice prepared from seven beetroot varieties grown in Upper Austria. *Journal of Food Composition and Analysis,* 42, 46-55.

Xie, F., Liu, W. C., Liu, P., Wang, J., Halley, P. J. & Yu, L. 2010. Starch thermal transitions comparatively studied by DSC and MTDSC. *Starch - Stärke,* 62, 350-357.

Index

G

H

I

K

L

M

N

O

P

Q

R

S

T

U

V

W